Storey's Guide to
RAISING
DUCKS

Dave Holderread

Storey Publishing

*The mission of Storey Publishing is to serve our customers
by publishing practical information that encourages personal independence
in harmony with the environment.*

Edited by Larry Shea and Deborah Burns
Cover design by Renelle Moser
Cover photograph by © Grant Heilman, Grant Heilman Photography;
  Giles Prett; PhotoDisc
Series design by Mark Tomasi
Text production by Susan Bernier and Deborah Daly
Photographs by Dave Holderread
Line drawings by Elayne Sears
Indexed by Peggy Holloway

Storey's Guide to Raising Ducks was previously published under the title *Raising
the Home Duck Flock*. This new edition has been expanded by 128 pages. All of the
information in the previous edition was reviewed and revised for this new text, which
offers the most comprehensive and up-to-date information available on raising ducks.

Printed in the United States by Versa Press
20 19 18 17 16 15 14 13 12 11 10 9

Library of Congress Cataloging-in-Publication Data

Holderread, Dave
    Storey's guide to raising ducks / Dave Holderread
        p.cm.
    Includes bibliographical references (p.).
    ISBN 1-58017-258-X (alk. paper)
    1. Ducks. I. Title.
SF505 .H66 2000
636.5'97—dc21   00-057341

CONTENTS

The Importance of Ducks — and This Book!

Ducks are the most underappreciated of all domesticated poultry and livestock species. It is hoped that this will change as more people learn about ducks, their characteristics, and their incredible productivity. This book provides wonderful coverage of both the basics and the details of duck production. Dave and Millie Holderread have spent many decades in working with ducks, and no other people are more qualified to pass along accurate and useful information about ducks.

Ducks, with their incredible array of different breeds, are important and useful resources for agriculture, yet also offer breeders much enjoyment. Each duck breed comes to us from a unique combination of place, people, and breeding goals, and each is well presented in this book.

Dave understands the intricacies of color genetics incredibly well, and has shared that wealth here. He also sheds light on the fine points of breeding ducks for high levels of production — both for eggs and for meat. The approaches in this book will ensure that ducks remain productive, hardy, beautiful, and enjoyable. Dave also clearly explains the finer points of selecting breeding birds for the production of great show birds.

The threat of losing distinct and productive breeds of ducks looms ever larger as the years go by. Conservation is ideally accomplished by lots of breeders raising lots of ducks, and selecting them for excellence in production, adaptability, soundness, and form. The approaches outlined here will enable the next generation of duck breeders to make great strides in the conservation of the genetic treasures that are contained in each breed package.

This book is a deep and useful resource for anyone from beginner to advanced breeder and is jam-packed with information available nowhere else. This includes recent scientific understandings and many insights and "tricks" from old-time breeders that are sure to contribute to success.

D. Philip Sponenberg, D.V.M., Ph.D.
Professor, Pathology and Genetics
Technical Coordinator, American Livestock Breeds Conservancy

ONE

WHY DUCKS?

The popularity of ducks is increasing in many areas of the world. It appears that the rest of us are beginning to understand what many Asians and Europeans have known for centuries — ducks are one of the most versatile and useful of all domestic fowl.

For the home poultry flock, we're looking for birds that produce meat and eggs efficiently, require a minimum of care and shelter, can find a good portion of their own food, destroy weed seeds, consume insects and other garden pests, are healthy and hardy, add beauty to our lives, and make good pets. For many situations, it's difficult to find a better all-purpose bird than the duck.

Easy to Raise

People who have kept all types of poultry generally agree that ducks are the easiest domestic birds to raise. Along with guinea fowl and geese, ducks are incredibly resistant to disease. While chickens usually must be vaccinated for communicable diseases and regularly treated for worms, coccidiosis, mites, and lice, duck keepers can normally forget about these inconveniences. Even when kept under less than ideal conditions, small duck flocks are seldom bothered by sickness or parasites.

Resistant to Cold, Wet, and Hot Weather

Mature waterfowl are practically immune to wet or cold weather and are better adapted to cope with these conditions than are chickens, turkeys, guineas, or quail. Thanks to their thick coats of well-oiled feathers, healthy ducks of most breeds can remain outside in the wettest weather. (Muscovies and any duck that has poor water repellency should have access to a dry shelter during wet weather.) While chickens have protruding combs and wattles that must be protected from frostbite, as well as bare faces that allow the escape of valuable body heat, ducks are more completely clothed and are able to remain comfortable — if they are provided dry bedding and protection from wind — even when the temperatures fall below 0°F. Ducks also do well in hot climates if they have access to plenty of shade and cool drinking water. During torrid weather, bathing water or misters can be beneficial.

Insect, Snail, and Slug Exterminators

Nurturing a special fondness for mosquito pupae, Japanese beetle larvae, potato beetles, grasshoppers, snails, and slugs, ducks are extremely effective in controlling these and other pests. In the midwestern United States, ducks are used to reduce plant and crop damage during severe grasshopper infestations. In areas plagued by liver flukes, ducks can help correct the problem by consuming the snails that host this troublesome livestock parasite.

General Comparison of Poultry

BIRD	RAISABILITY	DISEASE RESISTANCE	SPECIAL ADAPTATIONS
Coturnix Quail	Good	Good	Egg and meat production in extremely limited space.
Guinea Fowl	Fair–Good	Excellent	Gamy-flavored meat; insect control; alarm. Thrive in hot climates.
Pigeons	Good	Good	Message carriers; meat production in limited space. Quiet.
Chickens	Fair–Good	Fair–Good	Eggs; meat; natural mothers. Adapt to cages, houses, or range.
Turkeys	Poor–Fair	Poor–Fair	Heavy meat production.
Geese	Excellent	Excellent	Meat; feathers; lawn mowers; "watchdogs"; aquatic plant control. Cold, wet climates.
Ducks	Excellent	Excellent	Eggs; meat; feathers; insect, snail, slug, aquatic plant control. Cold, wet climates.

Under most conditions, two to six ducks per acre of land are needed to get rid of Japanese beetles, grasshoppers, snails, and slugs. To eliminate mosquito pupae and larvae from bodies of water, provide six to ten ducks for each acre of water surface. The breeds of ducks in the lightweight class and the larger bantam breeds are the most active foragers, making them the best exterminators for large areas. However, as noted in the chart of breed profiles on pages 20–21, most other breeds are good foragers.

Productive

Ducks are one of the most efficient producers of animal protein. Indian Runner and Campbell duck hens selected for egg production lay *better* than the best egg strains of chickens, averaging 275 to 325 eggs per hen per year. Furthermore, duck eggs are 20 to 35 percent larger than chicken eggs. Unfortunately, in many localities, strains of ducks that have been selected for top egg production are not as readily available as egg-bred chickens. Meat-type ducks that are raised in confinement are capable of converting 2.6 pounds of concentrated feed into 1 pound of bird. If allowed to forage where there is a good supply of natural foods, they can do considerably better. The only domestic animal commonly used for food that has better feed conversion is the broiler chicken, with a 1.9:1 ratio.

Excellent Foragers

Ducks are energetic foragers. Depending on the climate and the abundance of natural foods, they are capable of rustling 15 to 100 percent of their own food. Along with guineas and geese, ducks are the most efficient type of domestic poultry for the conversion of food resources that normally are wasted, such as insects, weed plants, and seeds, into edible human fare.

Aquatic Plant Control

Ducks are useful in controlling unwanted plants in ponds, lakes, and streams, improving conditions for many types of fish. In most situations, fifteen to thirty birds per acre of water are required to clean out heavy growths of green algae, duckweed (*Lemna*), pondweed (*Potamogeton*), widgeon grass (*Ruppia*), muskgrass (*Chara*), arrowhead (*Sagittaria*), wild celery (*Vallisneria*), and other plants that ducks consume. Once the plants are under control, eight to fifteen ducks per acre will usually keep the vegetation from taking over again.

In bodies of water containing plants submerged more than 2 feet, or when it is desirable to clean grass from banks, four to eight geese per acre of water surface should be used along with the ducks. (For geese to be effective, they must be confined to the pond and its banks with fencing.) Waterfowl are not effective in inhibiting the growth of tropical plants such as water lettuce and water hyacinth.

Garbage Disposal

Ducks will eat almost anything that comes out of the kitchen or root cellar. They relish many kinds of vegetable trimmings, table and garden leftovers, canning and fruit juice refuse, and most kinds of stale baked goods. To make it easier for these broad-billed fowl to eat firm vegetables and fruits, place apples, potatoes, beets, turnips, and such on an old board and crush them with your foot, or cut them into bite-sized pieces.

In a poultry program in Puerto Rico, we supplied a flock of forty Rouen ducks that were on pasture and had access to a 5-acre pond with nothing but leftovers from the school cafeteria. These garbage-fed birds remained in good flesh and showed no signs of poor health, although they produced 60 percent fewer eggs than a control group that was provided concentrated laying feed along with limited quantities of institutional victuals.

Useful Feathers

The down and small body feathers of ducks are valuable as filler for pillows and as lining for comforters and winter clothing. (See appendix E, Using Feathers and Down, page 296.)

Valuable Manure

A valuable by-product of raising ducks is manure. Duck manure is an excellent organic fertilizer that is high in nitrogen. In some Asian countries, duck flocks are herded through rice fields to eat insects and pick up stray kernels of grain. The birds are then put on ponds where their manure provides food for fish.

Gentle Dispositions

Rather shy, ducks are seldom aggressive toward humans. Of the larger domestic birds, they are the *least* likely to inflict injury on children or adults. In the 40 years I have worked with ducks, the only injuries I've sustained have been small blood blisters on my arms and hands — received while attempting to remove eggs from under broody hens — and an occasional scratch when the foot of a held bird escaped my grasp. If you do get scratched by the claws of a bird, prompt washing of the wound with hydrogen peroxide and the application of an antibiotic ointment will lessen the chance of infection.

A domestic Mallard hen with 10-day-old ducklings.

Decorative and Entertaining

Together with having many practical attributes, ducks are beautiful and fun to watch. A small flock of waterfowl can transform just another pond into a center of attraction and provide hours of entertainment.

I can still remember the first ducks I had as a young boy. Because our property had no natural body of water, I fashioned a small dirt pond in the center of the duck yard. After filling it with water, I watched as my two prized ducklings jumped in and indulged in their first swim. They played and splashed with such enthusiasm that it wasn't long before I was as wet as they were. And then — much to my delight — they began diving, with long seconds elapsing before they popped above the surface in an unexpected place. I was hooked, and continue to be intrigued by the playfulness, beauty, and grace of swimming waterfowl.

Comparison of Ducks and Chickens

CHARACTERISTIC	DUCKS	CHICKENS
Housing requirements	Minimal	Substantial
Height of fence to keep confined	2–3 feet	4–6 feet
Susceptibility to predators	High	Moderately high
Resistance to parasites and disease	Excellent	Fair
Resistance to hot weather	Good	Good
Resistance to cold weather	Good	Fair
Resistance to wet weather	Excellent	Poor
Foraging ability	Good–Excellent	Fair–Good
Scratching in dirt	None	Considerable
Probing in mud with bills or beaks	Considerable	None
Incubation period	28 days	21 days
Hatchability of eggs in incubators	65–90%	70–90%
Cost of day-olds in lots of 25	Ducklings double the price of chicks	
Starter feed — min. protein needed	16%	18%
Layer feed — min. protein needed	15%	15%
Age hens commence laying	16–24 weeks	18–24 weeks
Eggs laid per hen per year (wt.)	32–52 lbs.	22–34 lbs.
Feed to produce 1 pound of eggs	2.4–3.8 lbs.	2.8–4.0 lbs.
Part of diet hens can forage	10–25%	5–15%
Light required for top production	14–16 hours	14–17 hours
Annual mortality rate of hens	0–3%	5–25%
Efficient production life of hens	2–3 years	1–2 years
Protein content of eggs	13.3%	12.9%
Fat content of eggs	14.5%	11.5–12.5%
Cholesterol content of eggs	Ducks slightly higher	
Flavor of eggs	Similar; duck eggs sometimes stronger	
Feed to produce 1 pound of bird	2.5–3 lbs.	2–2.2 lbs.
Age to butchering	7–12 weeks	8–20 weeks
Typical plucking time	3–15 minutes	2–10 minutes
Color of meat	All dark	Light and dark
Protein content of flesh	21.4%	19.3%
Fat content of carcass	16–30%	5–25%
Usefulness of feathers and down	Excellent	Fair

Some Points to Consider

Despite the versatility and usefulness of ducks, there are several factors you should be aware of and understand. In some situations another type of fowl may prove more suitable.

Noise

Many people find the quacking of ducks an acceptable, if not pleasant, part of nature's choir. However, if you have close neighbors, the gabble of talkative hens may not be appreciated or tolerated. Some breeds (and some individuals within a breed) are noisier than others. Typically, Call ducks are the noisiest, with Pekins being the second most talkative. Under many circumstances, a small flock consisting of any other breed will be reasonably quiet if not frightened or disturbed frequently. Muscovies are nearly mute, making them the least noisy of all breeds. Also, drakes of all breeds have weak voices, and for the control of slugs, snails, and insects in town or suburb, they work fine.

Plucking

Most people find that plucking a duck is more time-consuming than defeathering a chicken. But then, most of us who have had the privilege of dining on roast duck agree that it is time well spent. Furthermore, duck feathers are much more useful than chicken plumes. With good technique and a little experience, it is possible to reduce the picking time to 5 minutes or less.

Pond Density

Having large numbers of ducks on small ponds or creeks encourages unhealthy conditions and can result in considerable damage to bodies of water. One of the feeding habits of ducks is to probe the mud around the water's edge for worms, roots, and other buried treasures. A high density of ducks will muddy the water and hasten bank erosion. On the other hand, a reasonable number of birds (15 to 25 per acre of water) will improve conditions for fish, will control aquatic plant growth and mosquitoes, and will not significantly increase bank erosion.

Gardens

Ducks do an amazing job of controlling slugs, snails, and various insects in gardens. However, to prevent the birds from doing more harm than good, the following guidelines must be observed:

1. Don't let birds in until the crops are well started and past the succulent stage.

2. Keep ducks out when irrigating or when the soil is wet.

3. Fence off tender crops such as lettuce, spinach, cabbage, and green beans.

4. Remove birds when low-growing berries and fruits are ripe.

5. Limit the number of ducks to two to four adults for each 500 to 1,000 square feet of garden space.

A method we have used successfully for decades is to pen ducks around the perimeter of the garden where they intercept migrating slugs, snails, and insects. Then, when we are working in the garden during the growing season, we allow a few ducks into the garden to "vacuum up" the pests they are so adept at ferreting out. When we are ready to leave the garden for the day, the ducks are enticed back to their garden-side enclosure with the feed can. During the nongardening season, the broad-billed exterminators are allowed into the garden daily.

Ducks can make a highly effective pest patrol in the garden, as long as you take care to keep them from tender plants and low-growing fruit.

Meat and Eggs

All types of poultry provide good food. However, there are variations in the flavor, texture, and composition of the meat and eggs produced by the diverse species. There are also differences in dietary needs and likes and dislikes of food among people. If you are seriously considering producing duck meat or eggs but have never cooked or eaten them, I recommend that you sample duck products before starting your flock. (This practice is wise before investing time and resources in any type of unfamiliar animal for food.) The following observations are presented to help you evaluate your first encounter with duck cuisine.

1. The flavor of meat and eggs from ducks whose diet included fish or fish products is often strong and not typical of good duck.

2. Ducklings that have been raised in close confinement and pushed for top growth (e.g., Long Island Ducklings) are much fatter than ducks that have foraged for some of their own food and have grown at a slower pace. The high fat content of quick-grown duckling makes its meat exceptionally succulent and provides valuable energy for persons who get strenuous physical exercise. However, many of us do not need the large quantity of fuel provided by fat ducklings.

3. Duck eggs typically have a slightly higher cholesterol content than the average chicken egg. As with fat in meat, the amount of cholesterol in eggs seems to be affected by the diet and lifestyle of the producing bird. Hens that are active and forage for a portion of their diet *may* produce eggs lower in cholesterol.

When eggs are eaten in moderation, the difference in cholesterol between duck and chicken eggs probably is insignificant for healthy people who get adequate exercise and eat sensibly — lots of high fiber and uncooked foods while going easy on the meats.

4. Duck eggs are excellent for general eating and baking purposes. When fried in an open pan, the whites of duck eggs are often firmer than those of chicken eggs. We prefer to steam-fry our duck eggs (see appendix D, Duck Recipes, page 294) for a softer-textured egg white. While it is often said that duck eggs are unsatisfactory for meringues and angel food cakes, we have not found this to be true. In fact, one of our favorite treats for special occasions is whole-wheat angel food cake made with duck egg whites (see page 295).

TWO

EXTERNAL FEATURES AND BEHAVIOR

Ducks are masterfully designed, down to the smallest detail, for an aquatic life. Being acquainted with the external features of ducks is a useful management tool and will deepen your appreciation for these waterfowl. I suggest that you take a few minutes to familiarize yourself with the accompanying nomenclature diagram.

Body Shape

For stability and minimal drag while swimming and diving, the lines of the body are smooth and streamlined; the underbody is wide and flat.

Feathers

Endowed with an extraordinary abundance of feathers, particularly on the underside of their bodies, ducks can swim in the coldest water and remain comfortable. Soft, insulating down feathers close to the skin are covered by larger contour feathers. Several times each day, ducks faithfully preen and oil their feathers. A water repellent is produced by a hidden oil gland at the base of the tail. As a duck preens, it frequently squeezes the nipple of the oil gland with its bill and spreads the excretion onto its feathers.

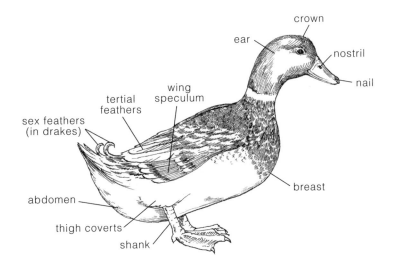

Wings

Ducks have long, pointed, rather narrow wings — except Muscovies, whose wings are broad and rounded. Most domestic breeds have lost their ability to fly, although Muscovies, Calls, East Indies, Australian Spotteds, and domesticated Mallards have retained their flying skills to varying degrees.

Tail

Muscovies have squarish tails that are 4 to 6 inches long. Other ducks possess shorter, pointed rudders. Excluding the Muscovy, mature drakes have several curled feathers in the center of their tails.

Bill

Long and broad, duck bills are well adapted for collecting food from water, catching flying insects, and rooting out underground morsels. With nostrils located near the head, ducks can dabble in shallow water and breathe at the same time. Due to hormonal changes in their bodies, duck hens with yellow or orange bills frequently develop dark spots or streaks on their bills when they begin to lay. These blemishes are not a sign of disease. Some waterfowl breeders believe they are an indication of high egg production.

With their long, broad bills, ducks are adept at catching flying insects, straining minute plants and animals from water, and probing in mud. The hooked nail at the end of the bill is used for nipping off tough roots and stems.

Eyes

The vision of ducks is much sharper than our own. Due to the location of their eyes, ducks can see nearly 360 degrees without moving their heads. This feature makes it possible for a feeding duck to keep a constant lookout for danger.

Feet and Legs

With webbed toes and short legs, ducks are not exceptionally fast on land, but are highly skilled swimmers, preferring to escape danger by way of water. The legs of ducks are injured more readily than those of chickens. Therefore it is advisable to pick them up by the neck and/or body.

Back

Normally a duck's back is moderately long with a slightly convex top line. In Rouens, a long back with a noticeable but moderate arch from the shoulders to the tail is a breed characteristic. However, ducks of all breeds with severely humped backs should be avoided for breeding purposes since they typically have poor fertility.

Keel

Ducks that are heavily fed and selected for large size and fast growth often develop keels. A keel is a fold of skin that hangs from the underbody and in extreme cases may run the entire length of the body and brush the ground as the bird walks. A well-developed keel is a breed characteristic of Aylesburies and Rouens, but is not preferred on any other breed.

Behavior

Just as you and I have distinct personalities, so each duck has its own peculiarities and habits. However, ducks follow behavioral patterns that you should understand if you're going to do a good job of raising them.

Pecking Order

This bird law regulates the peaceful coexistence of the duck flock. The number-one bird in the flock can peck or dominate all others, the number-two bird can dominate all but number one, the number-three bird can dominate all but numbers one and two, and on down the line until we reach the last individual who dominates no one.

If you introduce a new bird into an established flock, the existing pecking order is threatened, normally resulting in a power struggle that may evoke fighting. Unless the conflict is causing serious injury to the participants, you should not intervene, remembering that the roughhousing is necessary for the future peace of the flock.

Feeding

The natural diet of ducks consists of approximately 90 percent vegetable matter (seeds, berries, fruits, nuts, bulbs, roots, and succulent grasses) and 10 percent animal matter (insects, mosquito larvae, snails, slugs, leeches, worms, and an occasional small fish or tadpole). Sand and gravel serve as grinding stones in the gizzard. Ducks feed by dabbling and tipping up in shallow water, drilling in mud, and foraging on land. While ducks eat considerable quantities of tender grass, they're not true grazers as are geese.

Swimming

Skillful and enthusiastic swimmers from the day they hatch, ducks will spend many happy hours each day bathing and frolicking in water if it is available. However, ducks of all types and ages (particularly Muscovies and ducklings) can drown if their feathers become soaked and they are unable to climb out of the water. For this reason, ducks must be allowed to swim only where they will be able to exit easily. In general, it is safest (unless you are close by) to keep ducklings out of water until they are at least 2 to 4 weeks old. Ducks can be raised successfully without swimming water.

Mating

Ducks will pair off, although domesticated drakes normally mate indiscriminately with hens in a flock. If you raise several breeds and wish to hatch purebred offspring, each variety needs to be penned separately 3 weeks prior to and throughout the breeding season.

In single male matings, a drake can usually be given two to five hens, although males sometimes have favorites and may not mate with the others. In flock matings, one medium- or heavyweight-breed male should be allowed for every three to six hens, while one drake of the lightweight breeds should be provided for every four to seven hens. For good fertility, hens need to be bred by drakes at least once every 4 or 5 days. A few eggs that have been laid as long as 2 weeks after copulation may hatch.

Ducks prefer to mate on water, but most breeds can copulate successfully on land. Some large breeds, especially deep-keeled Aylesburies and Rouens, have higher fertility if they have access to water at least 6 inches deep.

Nesting

In the wild, a duck will hide her nest in a protected spot such as among tall grass or under a bush. Typically, the nest is a shallow depression in the ground that is lined with twigs, grass, leaves, and moss. If eggs are left in the nest for natural incubation, the duck will pluck down and feathers from her underbody to add insulation. Some ducks lay their eggs at random on the ground or even while swimming.

Fighting

In established flocks that are not overcrowded and have a proper sex ratio, ducks get along well together and rarely fight. If a new bird is introduced into a flock, there will be a short period of chasing, pushing, and wing slapping, but normally such conflicts subside quickly. Adult ducks seldom inflict injury on one another if there is not an excessive number of drakes in a flock. As a rule, drakes will not fight among themselves if there are no females around.

Life Expectancy

Ducks live a surprisingly long time when protected from accidental deaths. It is not unusual for ducks to live and reproduce for 6 to 8 years, and there are reports of exceptional birds living 15 years and longer. However, few ducks die of old age (except for those special pets!) since fertility and egg production often decrease after 3 to 5 years; hence, many owners do not consider it economical to keep ducks past this age.

Houdini Ducklings

Belknap Creek meandered through pastures and marshes near my childhood home. As I explored its marvels, one of my favorite activities was to hide in the tall grasses along the banks, watching wild ducks with their broods.

One day, as a Mallard and her seven newly hatched ducklings swam by, I couldn't resist jumping into the water and trying to catch them. As soon as I hit the water, the mother feigned a wing injury in an attempt to lure me away from her brood. The ducklings vanished. The muddied water soon cleared as I stood knee-deep in the creek. To my surprise, I saw several of the ducklings clinging to underwater vegetation, attempting to hide from me. I quickly climbed out of the water and concealed myself in the tall grass again. Within a few seconds, the ducklings bobbed to the surface and swam off under the watchful eye of their mother.

THREE

CHOOSING THE RIGHT DUCK

Choosing an appropriate breed can be enjoyable and plays an important role in the success or failure of your duck project. Unfortunately, novices often assume that a duck is a duck and acquire the first quacking, web-footed fowl they find. This mistake frequently results in expensive eggs or meat, needless problems, and a discouraged duck-keeper. Investing a little time at the outset in acquainting yourself with the basic characteristics, attributes, and weaknesses of the various breeds will go a long way toward eliminating unnecessary disappointments.

Important Considerations

The following questions are designed to help you identify your own special requirements when choosing a breed.

Purpose

What is your main reason for raising ducks? Is it for pets, eggs, meat, feathers, decoration, exhibition, insect and slug eradication, aquatic plant and algae control in ponds, or a combination of these and other aims?

Location

Where are you located? Some breeds are noisier than others — a fact that should be taken into consideration when you have neighbors in close proximity. Talkative breeds (Calls and Pekins in particular) also attract more predators.

Climate

What are your temperature extremes? All breeds are adaptable to an impressive range of climates. However, when the thermometer falls below 10 to 15°F, the Bantam breeds (especially Call and East Indie), the slender and tight-feathered Runner, and the bare-faced Muscovy can benefit from insulated housing (see chapter 14, page 191 for more on housing).

Management

How are you going to manage your ducks? Will they be locked up in a building, enclosed in a fenced yard, put on a pond, or allowed to roam freely? Will you provide all or most of their food, or only a small supplement to what they glean for themselves? Certain breeds adapt better than others to total confinement (especially Call, East Indie, Australian Spotted, Muscovy, and Pekin). Some breeds stay close to home while others will roam over a large area. Foraging ability also varies considerably. And due to small size or a trusting temperament, some breeds are more likely to be taken by predators.

Color

What plumage color is best suited to your situation? Color is significant for several practical reasons. The pinfeathers of light-plumaged birds are not as visible as those of birds with dark plumage, making it easier to obtain an attractive carcass with light-colored ducks when they are butchered. However, in many circumstances medium to dark birds are better camouflaged, making them less susceptible to predators. Also, if there is no bathing water available, colored ducks generally maintain a neater appearance than do white ones.

Preference

What do you like? Raising ducks should be enjoyable, so choose a breed that you find attractive and interesting. Read the following chapters on the various breeds, look at their pictures and choose one or more breeds that fit your tastes and needs. For most people and situations, there are several breeds that will work well. Some folks raise one breed to produce only eating eggs, another primarily for meat, and a third for natural incubation or decoration.

I'm frequently asked what my favorite breeds are. After working with every recognized breed and many that are not officially recognized, I can honestly say that I like them all.

Population Status

Do you wish to help save a rare breed? The population status of different breeds ranges from numerous to endangered. When you purchase and raise a rare breed, you're helping improve its chances of continued survival.

Useful Terminology

In order to read and speak intelligently about the different breeds of ducks, it is useful to understand some basic terminology.

- **Species.** The domestic ducks raised in North America belong to one of two species. Those that are descendants of the common wild Mallard go by the scientific name *Anas platyrhynchos domesticus*. The one breed that descended from the wild Muscovy is *Cairina moschata domesticus*. (No, you don't need to memorize these scientific names, unless you are working on an advanced degree in Quackology!)
- **Class.** The American Poultry Association has divided domestic ducks into four classes, based on size: Bantam, Lightweight, Mediumweight, and Heavyweight. When entering ducks in most shows, their classes must be identified on the entry forms.
- **Breed.** Over the years, waterfowl breeders in many parts of the world have developed distinctive types of ducks. As ducks in particular regions became uniform in physical characteristics, they were given names and recognized as separate breeds. Those breeds that have been effectively promoted and proven to have desirable traits have survived. A *recognized breed* is one that has been officially accepted into the Standard of one or more of the governing organizations of purebred poultry.
- **Breed type.** Most breeds can be identified by their unique type. *Type* refers to the shape and dimensions of the head, neck, and body, and the way these parts all fit together to form the overall silhouette that is characteristic of a breed. A bird is said to be *typey* if it possesses all of its breed's key characteristics for shape. (Some breeds, such as the Cayuga,

Breed Type

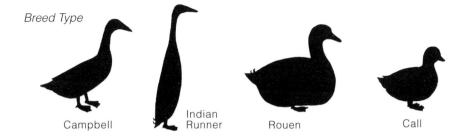

Campbell   Indian   Rouen   Call
                  Runner

Orpington, and Swedish, have similar size and type, and therefore are distinguished primarily by the color and pattern of their plumage.)

♦ **Variety.** Some breeds have more than one variety. In ducks, varieties almost always are distinguished from one another by the color or pattern of their plumage. An example of a breed that has just one variety is the White Aylesbury. On the other hand, Muscovies come in four recognized varieties: White, Black, Blue, and Chocolate. A *recognized variety* is one that has been officially accepted into the Standard of one or more of the governing organizations of purebred poultry.

Some breeds have *unrecognized varieties* that are created by hobbyists or commercial producers. These can become recognized if they go through the standardization procedure required by the American Poultry Association or the American Bantam Association (see chapter 8, page 92).

♦ **Strain.** Breeds consist of different strains. A strain is a group of birds that all descend from one flock or breeding farm, and are more closely related than the members of the breed at large. Strains are usually identified by using the name of their originator as a prefix; for example: Horton East Indies, Lundgren White Calls, Oakes Khaki Campbells, Sherraw Rouens, or Legarth Pekins.

When acquiring ducks, keep in mind that the strain is at least as important as the breed. For example, there are strains of Khaki Campbells that will consistently produce 300 to 340 eggs per female per year when managed properly. Conversely, there are other strains of Khaki Campbells that will barely lay 150 eggs per female per year. A broader example is the difference between production-bred and standard-bred strains as described in chapter 11.

♦ **American Poultry Association.** The APA is a membership organization that sanctions shows and publishes Standards for the recognized breeds and varieties of chickens, ducks, geese, and turkeys.

- ◆ **American Bantam Association.** The ABA is a membership organization that sanctions shows and publishes Standards for the breeds and varieties it recognizes in bantam chickens and ducks. The ABA and APA Standards do not always agree. See appendix H, page 302, for ABA contact information.
- ◆ **American Standard of Perfection.** This book contains a complete description of the ideal specimen in each breed and variety recognized by the APA. The *Standard* is periodically revised and updated and is available for the APA (see appendix H, page 302 for address). Most serious breeders and exhibitors own a copy. Local poultry clubs and public libraries sometimes have them available for loan.
- ◆ **Faults.** In purebred poultry, a fault is any characteristic of a bird that falls short of the ideal as described in the *Standard of Perfection*. It is frequently said that the perfect specimen has never been raised, and even the best birds have at least minor faults. Even among the most experienced and respected breeders and judges, complete agreement

Breed Profiles

WEIGHT CLASS	BREED	WEIGHT IN LBS. M / F		YEARLY EGG PRODUCTION
Bantam	Australian Spotted	2.2	2.0	50–125
	Call	1.6	1.4	25–75
	East Indie	1.8	1.5	25–75
	Mallard	2.5	2.2	25–100
Light	Bali	5.0	4.5	120–250
	Campbell	4.5	4.0	250–340
	Harlequin	5.5	5.0	240–330
	Magpie	6.0	5.5*	220–290
	Runner	4.5	4.0	150–300
Medium	Ancona	6.5	6.0	210–280
	Cayuga	8.0	7.0	100–150
	Crested	7.0	6.0	100–150
	Orpington	8.0	7.0	150–220
	Swedish	8.0	7.0	100–150
Heavy	Appleyard	9.0	8.0	200–270
	Aylesbury	10.0	9.0	35–125
	Muscovy	12.0	7.0	50–125
	Pekin	10.0	9.0	125–225
	Rouen	10.0	9.0	35–125
	Saxony	9.0	8.0	190–240

Note: Information presented in this profile is based on the average characteristics of each breed. Actual performance may vary considerably from the norm.

on the severity of faults does not often occur. What one person identifies as a significant fault, another may overlook.

Breeds of Ducks

Most domestic ducks raised in North America belong to one of twenty breeds or their hybrids. These water-loving fowl represent a marvelous cornucopia of sizes, shapes, and colors. By examining their breeding histories, we get an intriguing glimpse of the places, events, and traditions that have helped to shape the present duck world.

When reviewing the various breeds, always remember that birds within the same breed can vary greatly in their physical, practical, and personality traits. Furthermore, the environment they are raised in and the diet they consume can significantly alter not only their growth and productivity, but also their appearance. For descriptions of plumage colors and patterns, see chapter 10, Understanding Duck Colors.

EGG SIZE PER DOZEN (OUNCES)	MOTHERING ABILITY	FORAGING ABILITY	STATUS
20–24	Excellent	Excellent	Endangered
16–20	Excellent	Excellent	Common
18–24	Excellent	Excellent	Fairly common
24–28	Excellent	Excellent	Abundant
28–36	Poor–Fair	Excellent	Endangered
28–34	Poor–Fair	Excellent	Fairly common
29–34	Poor–Good	Excellent	Rare
30–38	Fair–Good	Excellent	Rare
28–36	Poor–Fair	Excellent	Common
30–38	Fair–Good	Excellent	Endangered
30–38	Fair–Good	Good	Common
30–38	Fair–Good	Good	Common
30–36	Fair–Good	Good	Fairly common
30–38	Fair–Good	Good	Fairly common
34–40	Fair–Good	Good	Rare
38–44	Poor–Fair	Fair	Rare
38–50	Fair–Excellent	Excellent	Abundant
36–46	Poor–Fair	Fair	Abundant
36–44	Poor–Good	Good	Common
36–46	Fair–Good	Good	Rare

*The APA Standard gives weight of 5 pounds and 4.5 pounds, respectively, for Magpie drakes and ducks, approximately 1 pound lower than typical.

FOUR

BANTAM BREEDS

The Bantam Class includes the miniature breeds of the domestic duck clan. These birds weigh between 18 and 40 ounces. They are popular for pets, decoration, and exhibition, but are also good seasonal layers and produce meat that has outstanding flavor and fine texture. Highly adaptable, they thrive in limited space under close confinement or can be allowed to roam at large where predators are not a problem.

Under many conditions, a patrol of eight to twenty of these little broadbills per acre of land provides an effective means of controlling many pests, including mosquito larvae, various insects, snails, and small- to medium-sized slugs. (Large slugs are best dealt with by larger ducks.) During the seasons of the year when there is an abundant supply of natural foods, bantam ducks that roam over a large area during the day require minimal supplemental feed.

Due to their small size, bantam ducks are more susceptible to nocturnal winged predators (such as Great Horned Owls) than are larger ducks. For this reason, in many localities it is prudent to pen bantam ducks in a covered enclosure from dusk to dawn. If left out at night on a pond, they have a better chance of surviving if there is an island with a covered area where they can hide.

Most bantam ducks are capable aviators, but in 40 years I have never had homegrown birds depart through the airways. However, do not expect them to stay at home if they are terrorized by predators (including dogs and people), frequently allowed to go hungry, or are overcrowded. When relocating them to a new home, it is safest to put bantams in a tightly fenced pen with a covered top until they are acclimated to their new residence. They can also be grounded by clipping the primary flight feathers of one wing.

Diminutive size is primarily genetically controlled. There is a tendency among bantams for some of the offspring to be larger than their parents. Therefore, to maintain the correct size, it is important to select small birds at each generation for breeding purposes. A common technique for enhancing the production of small ducks is to use linebred matings. This entails mating together related birds, such as first cousins, uncles, or aunts to nephews or nieces, or parents to offspring. If vigor and productivity decline due to excessive inbreeding, a carefully selected outcross (the mating together of birds that are not closely related) to another family or strain is required. Typically when you mate birds that are more distantly related, their offspring will be larger in size and more productive.

Nutrition and environment also affect the body size of bantam ducks. Ducklings that are grown out on a 15-percent protein ration that is balanced for all essential nutrients will be smaller than those provided a higher-protein diet. A feed regimen employed by some successful bantam breeders consists of giving ducklings chopped leafy greens (such as dandelions, leaf lettuce, or succulent young grass that has not been contaminated with chemicals) and chick-sized insoluble granite grit and feed made up of 1 part (by volume) uncooked rolled oats to 2 parts good-quality 18- to 20-percent protein waterfowl starter/grower crumbles. Ducklings that get out at an early age to forage and exercise extensively will generally not grow as large as those that are raised in close confinement and do little more than eat, drink, and sleep.

To maintain productivity from generation to generation, breed only from females that lay well-shaped eggs with good shells. To encourage good production and high hatchability of eggs, it is helpful to supply bantam breeding ducks with a high-quality waterfowl breeder or gamebird breeder ration containing approximately 20-percent protein during the breeding season (2 to 3 weeks prior to the first eggs, and throughout the laying period).

In single male matings, pairs or trios often produce the best fertility, although some drakes will successfully fertilize three or more ducks. In multiple male flock matings, it is often necessary to have at least two to three females per male to minimize fighting among the drakes. Most females are excellent natural mothers and sometimes are used for hatching the eggs of endangered wild ducks.

Calls

In the twenty-first century, this mini-duck is primarily raised for exhibition, decoration, and pets. The current pampered status of the Call is in stark contrast to its humble beginnings as a working, turncoat duck in Western Europe.

A direct descendant of the wild Mallard, the Call probably originated in Holland where it still goes by the name of Decoy Duck. Originally employed as live decoys, Calls were used to entice wild ducks to enter large funnel traps. Later, market hunters tethered Calls near their gunning stations to lure wild ducks to fly within shooting range. The Call may be the only breed of duck in history to have been selected for its voice. Their unique high-pitched voices (which carry over long distances) and ability to talk fast and with great persuasion made them invaluable sidekicks to generations of hunters.

In the early decades of its development, the physical appearance of the Call was hardly distinguished from its wild ancestors. Over the years, the bill was shortened and the body size reduced through selective breeding. Both the original Gray variety and the White sport were included in the first *Standard of Perfection* published by the American Poultry Association (APA) in 1874.

A white Call old drake with the plump body, short neck, "buffled" head, short bill, and horizontal carriage desired in this breed. A multiple winner bred by Gary and Kari Bennett of Oregon.

Since the middle of the twentieth century, Calls have made impressive strides in both popularity and quality. In North America today, this little charmer is a favorite among hobbyists. At poultry shows, Calls win more duck championships than any other breed.

Description

Modern Calls look like toy ducks. A tiny, plump, bowl-shaped body, short, broad bill not much larger than a thumbnail, stubby neck, large, round head that is wide across the skull, and short legs positioned near the center of the body are desired characteristics. Along with being the most talkative breed, they are also one of the most active. They do everything with vim and vigor — flirting, searching for tasty tidbits, bathing, preening, and flirting some more. Most Calls are easily tamed, and if talked to frequently will carry on animated conversations with their human caregivers.

Varieties

Calls are raised in nearly as many varieties as Runners, and new colors are being developed. Standard colors include Gray, White, White-Bibbed Blue, Snowy, Buff, Pastel, Blue & White Magpie, and Black & White Magpie. Nonstandard colors include Butterscotch, Saxony, Aleutian, Spot, Self-Blue, Self-Chocolate, Self-Black, White-Bibbed Black, Ancona, Penciled, Fawn & White, Khaki, Cinnamon, Dusky, Blue Dusky, Lilac, and Crested.

Selecting Breeders

When a breed is selected for extremely small size, it is especially important to choose breeding stock that is vigorous, active, strong-legged, and bright-eyed. Beginners often make the mistake of spending top dollar to buy extremely small Calls with the shortest bills they can find, only to discover that these specimens may be fine show birds, but are poor or non-producers.

Usually, your best bet is to select a small, typey male that displays abundant vigor, and mate him with one or more females that have excellent type and heads but are a bit larger in size so they have the capacity to produce viable eggs. If they are descendants of a strain that produces ducks with short bills, don't fret — breeding birds do not need to have extremely short bills themselves to produce offspring with excellent bills.

Selecting and Preparing Show Birds

The mantra of the Call exhibitor is "Everything else being equal, the smaller the size of the bird and the shorter the bill, the better." Most of today's top show specimens have bills measuring 1¼ inches or less in length. Also important are large, buffled heads with high crowns that rise abruptly from the base of the bill, full cheeks, bodies that are short and wide, horizontal body carriage, and good color.

Calls normally are such tidy birds that when provided with a balanced diet, clean pen, fresh bathing water, and plenty of shade, they will just about prepare themselves for shows. During the molt and for 6 to 8 weeks prior to a show, a ration that encourages excellent feather condition can be made by mixing 6 parts gamebird flight conditioner, 3 parts uncooked oatmeal or oat pellets, and 1 part cat kibbles. Chick-sized granite grit should be supplied free-choice, and greens such as lettuce or chard given at least three times weekly.

Comments

Calls have not been used extensively as decoys for nearly a century, but their loud, persistent talking is a reminder of their bygone vocation. Some people are charmed by the chatter; others find it annoying. Call eggs are usually white or tinted green or blue. Day-old Calls are often more delicate than the ducklings of other breeds. Special care should be taken to ensure that they are sufficiently warm and do not get wet. Once they are well started, Calls are typically quite hardy.

East Indies

The "little duck with many names" is an apt label for this widely admired breed. Brazilian, Buenos Aires, East Indian, East India, Emerald, and Labrador Duck are some of the more common names that have been or are still being used in some countries. In North America, however, its official name is East Indie.

Scant information is available concerning its formative years. Even though it is easy to imagine these emerald beauties at home on streams deep in tropical forests, their exotic names apparently have no correlation to their place of origin. It is generally accepted that the East Indie originated in

A young pair of Black East Indies. The drake won Reserve Champion Bantam Duck at the 1999 Northwest Winter Classic. Bred by Art Lundgren of New York, shown by Gene Bunting of Oregon.

North America during the first half of the 1800s, was then improved by British breeders (who still use the name East Indian) during the last half of that century, and was perfected to its current level of development by North American breeders during the last half of the 1900s.

Some of the early writers referred to the East Indie as a smaller and better-colored version of the Cayuga. In his book *The Practical Poultry Keeper* (1886), L. Wright wrote that the East Indian " . . . should be bred for exhibition as small as possible, never exceeding five and four pounds." As recently as the 1960s, 3-pound East Indies were common, though generally not preferred. During the 1990s, winning birds that I weighed tipped the scales at 16 to 20 ounces.

The East Indie was included in the first British *Book of Standards* in 1865, as well as in the first APA's *Standard of Perfection* in 1874. Today among hobbyists, this is the second most popular bantam duck breed, behind the Call. Currently East Indies are raised primarily for decoration, pets, and exhibition.

Description

Good East Indies are intermediate in type, positioned between the short, plump Call and the long, racy Mallard. Breed characteristics include a moderately small oval head, bold eyes, moderately long neck of medium diameter, medium-wide body that is moderately long, medium-length legs, and horizontal body carriage. The bill has a slightly concave top line when viewed from the side. Winning birds I measured during the late 1990s had average bill lengths of 1¾ inches in drakes and 1⅝ inches in ducks.

The iridescent green-black plumage of the Black East Indie must be seen before its beauty can be fully appreciated. When raising black ducks, keep in mind that some of them will have more or less white in their plumage, even if the breeding stock they are from has been carefully selected for generations.

Varieties

Black with green iridescence is the traditional color. Bill Mayer of Michigan developed Blue East Indies with fine type and small size, and introduced them to the public in 1999 at the Western Waterfowl Exposition in Albany, Oregon.

Selecting Breeders

Choose active, bright-eyed birds that have good breed type and small size. As with Calls, the tiniest birds are sometimes not productive.

Selecting and Preparing Show Birds

To be successful in good competition with East Indies, you must show birds that are small, have excellent type and color, show no white in their plumage (one small white feather can be the difference between a bird winning first or not placing), and are in first-class feather. Because brilliant black ducks normally have at least a bit of white in their plumage, it pays to go over a bird with a tweezers to remove any small white or faded feathers prior to taking them to a show. It is illegal to remove large wing and tail feathers. Some exhibitors wipe down their entrants with a silk cloth an hour or so prior to judging to enhance the feathers' sheen.

To improve the green sheen of the plumage, during the molt and for 6 to 8 weeks prior to a show, you can provide a ration consisting of 6 parts game-bird flight conditioner, 3 parts uncooked oatmeal or oat pellets, 1 part cat kibbles, and ½ part black-oil sunflower seeds. Chick-sized granite grit should be supplied free-choice, and greens such as lettuce or chard given at least three times weekly.

Comments

East Indies normally are not as talkative as Calls, although some strains have been crossed with Calls and carry the latter's distinctive voice. They can make great pets, but typically are a bit more shy than Calls. The first eggs of a laying cycle are often gray or black, gradually changing to lighter hues of gray, blue, or green as the season progresses. In my experience, East Indies are good natural mothers. If not overly inbred, the ducklings are hardy, especially after they are 1 to 2 weeks old.

Mallards

The wild Greenhead is native to most countries of the Northern Hemisphere and goes by the scientific name *Anas platyrhynchos platyrhynchos* (a-nas pla-ti-ring-kos). This much-admired species is believed to be the parent stock of all domestic duck breeds except for the Muscovy. No one knows for sure when Mallards were first domesticated, but there is evidence that Southeast Asians and Romans were raising ducks in captivity prior to 500 B.C.

Mallards are widely raised, primarily for hunting clubs, gourmet meat, decoration, pets, and exhibition. Interestingly, the APA did not include them in their *Standard of Perfection* until 1961.

Description

Authentic Mallards are elegantly streamlined. They have trim, teardrop-shaped bodies, slender necks that can telescope to surprising lengths, aerodynamic heads, and horizontal body carriage. A distinctive characteristic of the true Mallard is a long, slender bill. In good exhibition drakes, the bill measures 2¾ to 3 inches long, whereas those of ducks typically are 2¼ to 2½ inches long.

Most of the Mallards sold by hatcheries have been selected for larger body size and higher productivity. In the process, they have lost some of the elegance and refinement of their more authentic cousins.

Varieties

The Gray is the original wild color. Rare varieties include the Snowy, White, Golden, Pastel, and Blue Fawn.

This authentic Mallard duck, bred on our farm, displays her wing speculum and intricately penciled body plumage.

Selecting Breeders

Mallards readily adapt to domestication. In a short four or five generations of heavy feeding in captivity, they can increase in size and become less streamlined in type. If you wish to preserve their sleek characteristics, beautiful color, and outstanding foraging ability, it is necessary to select breeding stock in each generation that most closely resemble the wild birds.

Selecting and Preparing Show Birds

Small, streamlined Mallards with long bills and proper color compete successfully with the best of other breeds. In drakes, look for distinct white neck collar and dark chestnut breasts with a minimum of foreign color. In ducks, select medium-colored birds with distinctly penciled feathers. Mallards are naturally shy and normally require more cage training than other bantam breeds.

During the molt and for 6 to 8 weeks prior to a show, you can use the same ration as recommended for Call ducks to encourage excellent condition.

Comments

Along with Muscovy and Australian Spotted, the genuine Mallard is the most self-reliant of all breeds. In the right environment, females are outstanding mothers and will hatch a high percentage of their buff, green, or blue eggs. Mallard hens can be valuable for incubating the eggs of wild species and domestic breeds that are difficult to hatch in incubators.

In Canada, a migratory waterfowl permit is required to possess Mallards. In the United States, a permit is required to capture, gather eggs of, possess, or sell wild Mallards. In most states, a permit is not currently required to raise and sell domestic Mallards. However, regulations change from time to time so you should check with your local Fish and Wildlife bureau for current regulations. Mallards are not allowed in Hawaii at this time, because it is feared they will crossbreed with endangered wild ducks.

Australian Spotted

The charming and beautiful Australian Spotted, its "down under" name notwithstanding, originated in the United States. The late John C. Kriner Jr., of Orefield, Pennsylvania, in communications during the mid-1970s told me that he and Stanley Mason developed this breed (originally known as the Australian Spots) in the 1920s, and began exhibiting them in 1928. Kriner stated that the foundation stock was the Mallard, Call, Northern Pintail, and an unidentified wild Australian duck possessing spotted plumage.

The original birds were kept together and allowed to interbreed for several generations. The preferred offspring were then selected and bred from, thus forming the new breed. Because Pintail x Mallard crosses usually are sterile, I originally assumed that Pintails had not contributed to further generations. However, after raising and studying Australian Spotted for a decade and observing physical and behavioral characteristics normally not seen in pure Mallard derivatives, I am not as ready to dismiss Pintails as one of their ancestors. Every once in a while, a mule produced by a donkey x horse mating is fertile. Likewise it is possible that Kriner and Mason had a rare fertile Pintail x Mallard hybrid that did contribute genetic material to their new breed.

During a 1975 visit to the late Henry K. Miller's Blue Stream Farm located near Lebanon, Pennsylvania, Mr. Miller showed me his birds and told about developing his own strain of Australian Spotted in the 1940s. Originally this breed was classified and exhibited as a wild species.

Prior to 1990, this exquisite little duck was not readily available to the public. Thankfully its status today is improved, although the number of Australian Spotteds is still dangerously low.

Description

According to the originators, the Australian Spotted should be intermediate in type between the racy, long-billed Pintail and Mallard and the plump, short-billed Call. The bill is of medium length (1¾ to 2 inches long) and medium width. The head is oval, of medium size, and moderately streamlined, without the distinctive high forehead and puffy cheeks of the Call duck. The body is moderately racy and has a teardrop-shaped profile when viewed from both the side and top. The legs are attached near the center of the body, allowing for a nearly horizontal body carriage when the bird is relaxed. The Australian Spotted is a bit smaller than most wild Mallards, at 30 to 38 ounces.

The vocalizations of the Australian Spotted are not as high-pitched or plentiful as those of Calls.

A young pair of Bluehead Australian Spotted, bred on our farm, with excellent type and plumage markings. The light-colored underbody that runs up onto the chest of the drake is clearly visible.

Varieties

Greenhead is the original color. Today, Bluehead and Silverhead varieties are also being raised. The variety name refers to the color of the drake's head when in breeding plumage.

Selecting Breeders

Choose active, bright-eyed birds that have good breed type and moderately small size. Australian Spotted should not be as tiny as Calls or East Indies. The sides of the body and flanks of drakes should be strongly colored with chestnut-burgundy. Select ducks that have bold spotting.

Selecting and Preparing Show Birds

At this writing, the Australian Spotted is not yet in the APA's *Standard of Perfection*. They can be shown at most poultry shows, but they cannot compete against standardized breeds. In size and conformation, a good Australian Spotted looks like it could be a wild duck. Avoid birds that have strong Call characteristics. Australian Spotted are tidy birds and if you provide them with a balanced diet, clean pen with sufficient room, fresh bathing water, and plenty of shade, they, too, will prepare themselves for shows.

During the molt and for 6 to 8 weeks prior to a show, use the ration recommended for Calls and Mallards to encourage excellent condition.

Comments

Along with their diminutive size and delightful plumage, Australian Spotted have proven to be personable and long-lived. They are exceptionally hardy and outstanding foragers, and the females are the best layers of the Bantam breeds. Their cream, blue, or green eggs hatch into lively ducklings that display a tremendous zest for life. They are the quickest maturing breed I have observed. Sometimes, 3- to 4-week-old drakelets engage in courtship displays that I have not observed in other breeds until at least twice that age. We have had ducks commence laying at just over 3 months of age.

Because of the rarity of the Australian Spotted, their future is tenuous. They merit being saved, not only for their wonderful aesthetics and personality, but also for their many practical attributes.

FIVE

LIGHTWEIGHT BREEDS

Ducks in the Lightweight Class weigh between 3½ and 5½ pounds. They are good to outstanding egg layers. Although classified as nonsetters, individual birds that will successfully incubate and hatch eggs can be found within all of these breeds. If managed properly, each yields gourmet meat in a portion suitable for two to three people.

The lightweight breeds are the most active foragers among ducks, covering more ground as they eagerly search and consume slugs, snails, insects, and other edibles. Keepers of large livestock find that ducks are effective agents for eliminating liver fluke infestations. While all ducks enjoy swimming water, these breeds are more at home on land than most. Under most circumstances, they are not capable of sustained flight. However, they can propel themselves over 2- to 3-foot-high barriers if startled or hungry. As a group, they tend to be high-strung, although there is wide variation from strain to strain and even bird to bird. If dealt with calmly, they normally become friendly.

Drakes in this class tend to have high libido. In single male matings, a drake will typically fertilize anywhere from four to six females. In flock matings, five to seven ducks per drake normally gives good results. To avoid injury to the females, it is important not to have an excessive number of males.

Laying ability and egg size are characteristics that are strongly influenced by the father. Therefore, in order to maintain high yields of large eggs, it is prudent to carefully choose breeding drakes from high-producing families. If efficient egg production is one of your principal goals, one of the Lightweight breeds should be given top priority.

Bali

The Bali is one of the oldest domestic ducks. The early histories of the Bali and Runner are most likely parallel since the former appears to be essentially a crested version of the latter. Duck herding has been an integral part of traditional rural life on Bali and other Indonesian islands and in various parts of the Southeast Asia mainland for thousands of years. In duck flocks kept in these areas, it was common for crested birds to be among their numbers. In North America and Europe, the crested birds became known as the Bali, Balinese Crested, or Crested Runner; those with plain heads were simply called Runners. The Bali is not recognized by the American Poultry Association (APA) and is only sporadically available.

Description

The main differences between the Bali and Runner are the former has a chunkier body, is a bit wider across the shoulders, has a coarser head and bill, and has a crest on its head. Standing at attention, typical body carriage is 60 to 75 degrees above horizontal. All other characteristics are similar to those of Runners.

Due to the genetic makeup of crested ducks, when mated together one out of three offspring on average will be plainheaded and two will have crests. Another characteristic is that ducklings out of the same parents typically have crests that vary widely in size and shape. Crested ducks are prone to produce some offspring with arched necks, deformed backs, or balance problems.

A White Bali drake displaying typical breed characteristics of a rounder head, thicker neck, and chunkier body than its close relative, the Runner.

Varieties

Bali have been bred in many of the same variations as Runners. However, the most common colors are white, various shades of brown, and a modified Mallard pattern.

Selecting Breeders

First and foremost, select birds that have strong legs and an abundance of vitality. Do not breed from birds with balance problems, deformed backs, or severely kinked necks. Because arched necks are so prevalent in crested ducks, breeding from such specimens is sometimes unavoidable.

When possible, choose birds with smooth, well-rounded crests that are attached as high on the head as possible. For breeding birds, the size of the crest is not critical. In fact, those with smaller crests are usually more productive and will normally produce some offspring with significantly larger headgear.

The way to produce the largest number of offspring with the fewest deformities is to mate crested to noncrested birds. This mating, which was the usual procedure in the homeland of the Bali, typically produces half crested and half noncrested ducklings. Avoid breeding from birds with short or dished (concave) bills, round heads, and thick, heavyset bodies with low body carriage.

Selecting and Preparing Show Birds

Look for graceful, moderately slender, upstanding birds with good height, straight backs, and medium-sized crests that are as balanced and round as possible. Because Bali tend to be shy, it is helpful to talk to them reassuringly at feeding time and to cage-train them prior to taking them to their first show. When transporting them to shows, use containers that are taller than the birds to avoid temporary kinked necks. During the molt and for 6 to 8 weeks prior to a show, a good feeding regime for encouraging excellent condition without excessive weight is to supply oats (whole or pelleted) and granite grit free-choice. In addition, feed a mix of 7 parts gamebird flight conditioner and 1 part cat kibbles morning and night in a quantity that they will clean up in 5 minutes.

Comments

In my opinion, the Bali is the most unusual of all breeds. With their dignified yet comical appearance, they can bring a smile to the face of even the most serious-minded person. They are generally good layers of white, green, or blue eggs. Locating specimens to get started with is a worthwhile task, as Bali is a fascinating breed to raise.

Campbells

Mrs. Adele Campbell of Uley, Gloucestershire, England, originated this breed in the late 1800s. In a book titled *Ducks — Show and Utility*, C. A. House included the following information on the early history of the breed:

> Writing to me on the origin of the breed, Mrs. Campbell says: "The real beginning of the Khaki Campbell was a great appetite on the part of my husband and son for roast duckling! To make ducklings, one must have duck eggs, and I had just one duck, a Fawn & White Indian Runner*, which laid 195 eggs in 197 days. She was the only duck in the yard, a rather poor specimen in appearance, and no pedigree. However, I thought some good layers might be expected from her, but I wanted a little more size and mated her to a Rouen drake. . . . The original Campbells were practically this cross except that one season a Mallard drake was used. . . ."

From this humble foundation, one of the world's most prolific egg laying breeds was developed.

Introduced to the public in 1898, the original Campbells resembled poorly colored Mallards. Drakes had dark green heads, ill-defined neck rings, grayish bodies, and pale claret breasts. Ducks were grayish-brown, penciled with dark brown, often with a patch of white on the neck.

In an attempt to make a buff-colored duck, Mrs. Campbell mated her original Campbells back to Penciled Runners. Rather than the sought-after buff-plumaged birds, she ended up with tannish-brown ducks that reminded

---

The variety called Fawn & White in Great Britain is known as Penciled in North America. The American Fawn & White is a different color and is not recognized in England.

her of the khaki-colored uniforms worn by the British soldiers fighting in the South African War of 1899. The Khaki variety, which was introduced in 1901, grew in popularity and the original color gradually disappeared.

The ability of the Campbell to produce prodigious quantities of eggs was evident early in its history. Prior to 1920, reports had surfaced in British poultry magazines and books of small flocks of Campbells that averaged close to 300 eggs per duck per year. The record for an individual duck of that era was a mind-boggling 360 eggs laid in 365 days.

The most successful large-scale breeder in the history of Campbells was Mr. Aalt Jansen of Ermelo, Holland. Mr. Jansen was a fisherman who began raising Khaki Campbells in 1921. By 1950, the famous Jansen's Duck Breeding Farm had approximately 50,000 birds that averaged 335 to 340 eggs per duck at 17 months of age. One of their amazing Campbells began laying at 108 days of age and laid an egg a day for 405 consecutive days. The Jansens' success was based on their pedigreed breeding program, carefully formulated feeds, and outstanding farm management.

The earliest record that I have found of Campbells in the United States is the importation from England of nine birds in 1929 by Perry Fish of Syracuse, New York. The Khaki Campbell was admitted to the *American Standard of Perfection* in 1941.

After languishing in relative obscurity in North America, Campbells began to be raised in greater numbers during the late 1970s. The primary causes for this upswing were threefold: the back-to-the-land movement that started in the 1960s, the surge of Asian immigrants (many of whom relished duck eggs) at the end of the Vietnam War in 1975, and the importation of high-producing Campbells from the Kortlang Duck Farm of England in 1977.

Today, most Campbells continue to be raised for their wonderful practical qualities, although fine exhibition specimens can be seen at many of the larger poultry shows.

Description

In conformation, this is a breed of moderation. The late Henry K. Miller of Pennsylvania, a longtime Campbell breeder and respected poultry judge, told me on a visit to his farm, "Campbells are built the way a duck should be built." They are active, moderately streamlined birds. The head, bill, neck, and body are all modestly long, with a sprightly body carriage of 20 to 40

An old pair of egg-bred Khaki Campbells from our farm possessing typical markings and type. This duck laid 1,182 eggs in 6 years.

degrees above horizontal. When in prime feather condition, the warm earth tones of the Khaki make it a lovely duck to look at, especially when seen on a carpet of green grass.

A quirk in the genetic makeup of Khaki and Dark Campbells results in some of the offspring hatching with a bit of white on the front of the neck and/or under the bill, even if only solid-colored birds have been carefully selected for breeding stock for many generations.

Varieties

Khaki is the main variety. Whites were derived as a "sport" from the Khaki and are raised in small numbers both in England and North America. The Dark Campbell, which exists in North America but is rare, was originated by Mr. H. R. S. Humphrey of Devon, England, in order to make possible the production of sex-linked Campbell ducklings. Pied Campbells, which have the same pattern as Penciled Runners, were bred out of pure Campbell stock in Oregon in the early 1980s.

Selecting Breeders

To maintain high production in laying ducks, it is critical to choose breeding birds that are robust, bright-eyed, and have strong legs. Drakes should be sons of prolific dams that produce strong-shelled eggs of the desired size.

In the Northern Hemisphere, especially above the 30 degrees parallel, the normal time for birds of most species to lay is during the spring when both daylight hours and temperatures are increasing. As Leslie Bonnet stated succinctly in his book *Practical Duck-keeping* (1960), "In the spring even sparrows lay." Therefore, it is prudent to select breeder ducks that lay out of season, especially during the months when daylight hours are the shortest and the weather the most inclement. Experience has shown that if you breed only from ducks that are the best out-of-season layers, annual egg production will increase from one generation to the next until their genetic potential has been reached.

Longevity and lifetime egg production are valuable characteristics to select for. In our breeding program, we prefer to use ducks that have sustained high egg yields for 4 or more years. It is not unusual for these females to have lifetime production records of more than 1,200 eggs. Many of our best annual and lifetime layers have been produced by two-way or four-way line crosses. (A line is a carefully inbred family of birds.) When two or more inbred lines of ducks are crossed, their offspring typically have improved production.

Some sources have stated that ducks with certain physical characteristics (such as long bills with level top lines, eyes that are set high in the head, and long bodies) are the best layers. After raising Campbells for more than three decades, I have not found this to be true. Some of our best layers (including a duck that laid 357 eggs her first year of production and went on to produce many excellent sires and dams) possess average-length bills with concave top lines, eyes that appeared to be set relatively low in the head, and medium-length bodies. In my experience, the laying potential of a duck cannot be reliably judged by external characteristics.

Selecting and Preparing Show Birds

Most judges prefer sleek birds with moderately long bodies, nearly straight necks of medium length, refined heads, and medium-length bills with a nearly level top line. Body carriage should be intermediate between the upright Runner and the horizontal Call (about 30 degrees). Solid-colored varieties should have no white feathers visible when they are in the show cage and must not have clear yellow bills. Some judges also check along the jawline and under the bill for white feathers.

Because khaki plumage is susceptible to fading, successful exhibitors often keep their birds in the shade when the sun's rays are the strongest. Campbells tend to be high-strung, so cage-training them prior to their first exhibition often helps them show to their best advantage. During the molt and 6 to 8 weeks prior to a show, you can use the same feeding recommendations as outlined above for Bali ducks to encourage excellent condition without excessive weight.

Comments

If they consume an adequate diet, are kept calm, provided sufficient space, and run in flocks consisting of no more than an ideal of 50 to a maximum of 200 birds, Campbells have proved to be amazingly adaptable. They have performed admirably in environments ranging from arid deserts with temperatures of 100°F to humid tropical rainforests with more than 200 inches of annual precipitation to cold Northern regions where temperatures can remain below 0°F for weeks at a time.

A Word of Caution

If you want good egg yields, make certain you acquire authentic Campbells that have been selected for high egg production. The laying ability of birds in some flocks has been allowed to deteriorate and, much too often, crossbreds are being sold as Campbells. Any alleged Campbell that has facial stripes or weighs more than 6 pounds is a crossbred.

Egg-bred Campbells are the best layers of all recognized duck breeds, and in many environments are the most proficient egg producers of all avian species. Campbells lay eggs with superb texture and flavor and usually with pearly white shells. However, an occasional bird of authentic bloodlines will lay eggs with blue or green shells. For the home flock, I consider Campbells to be one of the best of all birds for the efficient production of eating eggs.

Welsh Harlequin

Leslie Bonnet, a commercial duck breeder who lived near Criccieth, Wales, originated the Welsh Harlequin in 1949 from two light-colored ducklings hatched out of pure Khaki Campbells. For the next 30 years, Mr. Bonnet distributed Harlequins throughout the world. The color and conformation of the Harlequin changed over the years, suggesting that new blood was introduced.

John Fugate, of Tennessee, imported hatching eggs from Bonnet in 1968. By 1980, the descendants of the original imports were confined to two small flocks. To augment the gene pool, adults were imported from Great Britain in 1981. Starting in 1984, Harlequins were made available to the public in North America and are currently fairly well established. Several British waterfowl judges who have visited the United States in recent years have commented that the Harlequins being bred here are more authentic than many of those found in their homeland.

An old pair of silver phase Welsh Harlequins bred on our farm with typical type and color markings. This duck's plumage is shaded with soft fawn.

Harlequins are primarily raised for their wonderful practical attributes. However, they are stunningly beautiful ducks that are highly decorative and make dazzling show birds. Currently, they are not in the *American Standard of Perfection*, but their qualifying meet for admission into the APA is tentatively planned for 2001.

Description

In conformation, the Harlequin closely resembles the Campbell. They are moderately streamlined throughout, with relatively long bodies, medium-width backs, nicely rounded chests, smooth underlines, and sufficient width between their legs to allow good capacity in the moderately full abdomens. Their necks are of medium length and held nearly vertical. Heads are trim and oval-shaped with medium-long bills that are only slightly concave along their top line. The Harlequin body carriage is a sprightly 20 to 35 degrees above horizontal, augmenting their active foraging habits. The resplendent colors and pattern of the Harlequin make it difficult to comprehend that the genetic differences between their plumage and that of the Campbell is the result of a single pair of genes.

Varieties

The original color of the Harlequin was the Golden, which has no black pigment and is equivalent to Khaki in Campbells. The Golden phase has soft colors and is rare in North America and possibly extinct in Great Britain. The Silver variety apparently arose at least 10 years after the origin of the breed, and has the same relationship to the Golden phase as Dark has to Khaki in Campbells. Silver-phase birds have more contrast and brilliance in their plumage and are by far the most common today.

Selecting Breeders

Birds used for breeders should be robust, strong-legged, free of physical deformities, heavy layers, and of correct body type and color. To help perpetuate the authentic Harlequin, avoid the following characteristics: more than a half pound above or below typical weights; short, blocky bodies; large, coarse heads; distinct Mallard-like facial stripes; light-colored bills in ducks; and poor producers.

Selecting and Preparing Show Birds

In size, type, and stance, the ideal show Harlequin looks like a slightly longer and larger Campbell. The size difference between these two breeds is approximately a half pound, which is a 10-percent difference. Cage-training Harlequins prior to their first exhibition can help them show to their best advantage. During the molt and for 6 to 8 weeks prior to a show, you can use the same feeding recommendations as outlined for Bali to encourage excellent condition without excessive weight.

Comments

When we took our first tray of Harlequin ducklings out of the hatcher in 1983, there were several distinct bill colors. Upon vent sexing the hatchlings, it was discovered that those with the darker bills were almost all males and those with light-colored bills ending in a dark tip were females. Mr. Bonnet was contacted and he replied that he was unaware of any sex-linked bill color in Harlequin ducklings. However, after vent sexing thousands of Harlequin ducklings during the subsequent years, I have found that they can be sexed with at least 90-percent accuracy by day-old bill color. A few days after hatching, the distinction in bill color begins to disappear.

Harlequins have proved to be one of the most important additions to the North American duck roster in the past 50 years. They are highly adaptable, outstanding layers, active foragers, excellent producers of lean meat, beautifully colored, and, due to their light under-color, they pluck almost as cleanly as white birds when dressed for meat. Even though their egg production is near or equal to bred-to-lay Campbells, a fair number of the females will successfully incubate and hatch eggs. Most Harlequins lay eggs with pearly white shells, although a minority lay colored eggs. Harlequins richly deserve their growing popularity.

Magpie

Oliver Drake and M. C. Gower-Williams of Wales are credited with developing this charming duck. Magpies are not mentioned in classic works published prior to the early 1900s, indicating that either they were not widely known or did not yet exist. From its size, conformation, and the genetic composition of its plumage pattern, Magpies almost certainly have Runners in their ancestry.

A Black and White Magpie old drake with exceptional plumage markings. Bred on our farm, it was the winner of Reserve Champion Light Duck at the 1993 APA National.

One interesting possibility is that the Magpies descended from the Huttegem, an old Belgium duck that was raised in great numbers during the 1800s in the duck growing district near Ondenaarde, 30 miles west of Brussels. The Huttegem is thought to have been developed from the Termonde and Runner breeds. Old pictures of the Blue Huttegem clearly show birds that genetically carry white bib and Runner pattern genes, which means they would have produced some offspring with classic Magpie markings. Furthermore, in Edward Brown's *Poultry Breeding and Production* (1929), his shape description of the Huttegem's head, bill, body, and station is an amazingly accurate portrayal of the Magpie.

According to Darrel Sherraw's book *Successful Duck and Goose Raising* (1975), Magpies were imported from Great Britain in 1963 by the late Isaac R. Hunter of Michigan. Following this importation, a handful of breeders, Mr. Sherraw, Jim Cleaver, and Curtis Oakes, all of Pennsylvania, and the Urch family of Minnesota, kept the Magpie alive in North America. In 1977, the Magpie received a boost when it was admitted to the *American Standard of Perfection*. Starting in 1984, Magpies became more readily available and since then have been gradually increasing in numbers.

A factor that hindered the popularity of Magpies among hobbyists was the virtual impossibility of meeting the specifications for bill and leg color as written in both the British and American Standards. In 1998, the APA revised its Standard specifications to bring them in line with the genetic potential of the Magpie, which should encourage more specialty breeders to take up this worthy breed.

Description

The name of this breed refers to its distinctive markings. In the ideal specimen, the plumage is predominantly white, offset by two colored areas — the back (from the shoulders to the tail) and the crown of the head. Breeding well-marked Magpies is a constant but highly rewarding challenge. Even out of the best matings, only some of the offspring will have good exhibition markings. However, a great advantage of the Magpie is that the best-marked ducklings can be identified at hatching. The best-marked birds can be raised for exhibition while the others can be raised for breeding, pets, egg production, or meat. An interesting characteristic of Magpies is that as they age, the mantle and cap, especially in ducks, will typically become mottled with white and may eventually turn totally white.

In conformation and station, the Magpie closely resembles the Campbell and Harlequin. The American weight specification for old drakes is 5 pounds, whereas in Great Britain it is 5½ to 7 pounds. American-bred Magpies vary considerably in size and anyone who prefers a larger bird could select a strain that would weigh 6 to 7 pounds. Due to their white underbody, they dress off almost as cleanly as solid white ducks.

Varieties

Black and Blue are Standard colors. Silvers are a natural product of Blues; Chocolates are extremely rare.

Selecting Breeders

A mistake most beginners make with Magpies is discarding many of the birds that make the best breeders. Magpies are a breed in which mating together the best show birds usually does not produce the best-colored offspring. (See chapter 10 for details on mating Magpies for color.)

Selecting and Preparing Show Birds

Good show birds have caps that cover at least half of the crown of the head and a clearly defined back mantle. Because the Magpie pattern is a challenge to perfect, the main emphasis by both exhibitors and judges should be clearly defined markings that are symmetrical rather than the exact size or shape of the markings. The bill is yellow or orange in young birds, gradually turning green with age. A long, sleek body with a nicely rounded chest, good width between the legs, and a body carriage of approximately 25 degrees above horizontal are desired. A moderately long neck with a racy head and bill finish off the ideal show bird.

Cage-training Magpies prior to their first exhibition can help them show to their best advantage. As with all marked varieties of show birds, most Magpies benefit from the removal of stray off-colored small feathers in the head, neck, and body plumage a day or two before a show. However, the removal of large tail or wing feathers is not permitted. During the molt and for 6 to 8 weeks prior to a show, you can use the same feeding recommendations as outlined for Bali to encourage excellent condition without excessive weight.

Comments

In 1987, we sent 500 ducklings in six breeds to Guatemala. These birds were distributed to Ketchi Indian villagers to see which breeds would perform the best under their subsistence farming conditions. Although all of the breeds tested performed well, the Magpies were the Ketchis' favorites due to their all-around performance and eye-catching markings.

Magpies are a true triple-duty duck. Along with being highly decorative, they are wonderful layers of large eggs and produce gourmet meat. Everyone who raises them can have the satisfaction of knowing they are helping perpetuate one of the rarest standard breeds of ducks.

Runners

Modern Runners are descendants of the traditional herding ducks raised along the coastal regions of the Indo-Chinese Peninsula and the islands of Southeast Asia. Hieroglyphics have been found in ancient Javan temples that suggest Runner-type ducks have existed for more than 2,000 years.

Raising and herding ducks has been a traditional lifestyle in parts of Asia for many centuries. One form of duck herding, which is still practiced, consists of herders taking their ducks out to rice paddies and fields during the day where the birds glean shattered grain, weed seeds, snails, insects, larvae, small reptiles, and such. The herder may carry a long pole, often adorned with a flag or some feathers, which the ducks recognize. When their destination is reached, the pole is stuck in the ground, and the ducks learn to stay within sight of the guidepost.

At day's end, the herders either lead or follow their charges home, where the ducks are typically enclosed in a bamboo or clay encampment. Because this type of duck normally lays in the early morning hours, their eggs are easily gathered prior to the foray into the countryside. Over their long history of being herded, a duck evolved that was a deft forager and could travel long distances at a quick pace.

Tradition has it that Runners were introduced into the United Kingdom from Malaya by a ship's captain around 1850. These birds landed at Whitehaven on the northwest coast of England, and were bred by farming friends of the captain in Cumberland County. They were blue with white markings and brown with white, the latter probably forerunners to our modern Penciled variety.

Before long they had spread to farmers in the adjoining county of Dumfries in Scotland. According to the *British Waterfowl Standards* (1982), fawn-colored Runners were shown at the Dumfries Show in 1876, Penciled were shown in 1896, and the Black and Chocolate varieties were included in the written *Standard* of 1926.

In North America, after considerable controversy over what color "true" Runners were, the Fawn & White variety was admitted to the American Standard in 1898. Finally, after cooler heads prevailed and it was recognized that there could be more than one color of Runners, both the pure White and Penciled varieties were standardized in 1914. The Black, Buff, Chocolate, Cumberland Blue, and Gray varieties were admitted to the American Standard in 1977.

Due to their unique appearance and high egg production, a Runner boom swept across Great Britain, France, Germany, and North America. Books and hundreds of articles were written on Runners, and their laying prowess was widely advertised and often exaggerated. Demand was so high that crossbreeding was common and any webfoot remotely resembling a Runner was peddled as the real article. Authentic Runners were saved by a few dedicated breeders and by adding new blood with direct imports from Asia as late as 1924. The boom inevitably dissipated, but Runners remain popular due to their practical qualities and unique appearance. At many shows, Runners constitute one of the largest classes of ducks and win their fair share of top awards.

Description

Of all the domestic ducks, the Runner departs the far-thest from its Mallard ances-tors in type and station. Runners are a splendid ex-ample of what selective breeding can accomplish. Proper Runner type is often graphically described as a

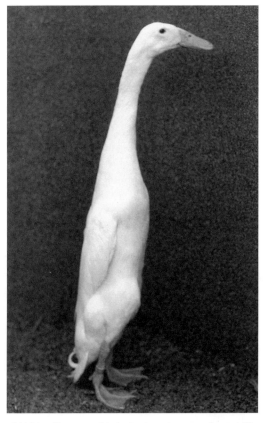

A White Runner old drake bred on our farm with the classic wine-bottle shape. It won Best Bird in Show at the Washington Feather Fanciers 1994 Autumn Classic.

wine bottle with a head and legs. When relaxed and strolling about their home range, Runners typically have a posture that is 45 to 75 degrees above horizontal. When startled or standing at attention, good specimens will strike a nearly perpendicular pose. Perched on top of a long, slender, and straight neck is a slim, wedge-shaped head and bill. The eyes are set higher in the head than in any other breed. When excited, good Runners will hold their tails straight down, in line with their back. When relaxed, even the best show birds normally hold their tails somewhat elevated. In contrast to the waddle of most ducks, a classic trait of the Runner is a smooth running gait.

Varieties

Runners have more varieties than any other breed. Standard varieties include Fawn & White, Penciled, White, Black, Buff, Chocolate, Cumberland Blue, and Gray. Nonstandard varieties include Fairy Fawn (one of the original varieties from Asia, it resembles muted Mallard colors), Blue Fairy Fawn, Golden, Saxony, Blue Fawn, Pastel, Trout, Dusky, Khaki, Cinnamon, Silver, Lavender, Lilac, Blue-Brown Penciled, Blue-Fawn Penciled, Emery Penciled, Porcelain Penciled, and Splashed. New varieties continue to be developed.

Selecting Breeders

In all breeds, there is a tendency for some offspring in each generation to regress slightly toward the original type of their wild ancestors. Therefore, to perpetuate the classic characteristics of the Runner, breeding stock must be carefully chosen.

Faults to avoid include low body station; short, stocky bodies with prominent shoulders and chests; short necks; round heads with prominent foreheads; short bills and/or concave top line; and tails that are constantly cocked upward, even when the bird is excited. (Many Runners will hold their tails down better in the unfamiliar surroundings of a show than they do at home.) Because bills want to regress to the original concave shape, it is useful to select some breeders that have powerful bills with a slightly convex top line.

Look for strong legs and a smooth running gait. A bird that does not move about freely under normal circumstances should not be used for breeding purposes. Keep in mind that any duck that is terrified or run too hard will likely go down in its legs temporarily.

Selecting and Preparing Show Birds

Many of the better shows have Runners judged in a ring rather than cages. This method has the advantage of allowing the running gait of the contestants to be evaluated, eliminating any possibility that a bird with weak legs will win. Some Runners pose the best in cages, but others show off to their best advantage in the ring. Keep in mind that some Runners that look quite average at home show off like champs when put in the environment of the exhibition hall. Until a bird has been shown several times, it can be difficult to evaluate its true worth.

An outstanding exhibition Runner is smooth, slender, and nearly vertical, with an imaginary straight line running from the back of the head through the neck and body to the tip of the tail. Everything else being equal, the taller the bird and the longer and straighter its bill, the better.

Good Runners are genetically slender and do not have to be fed dangerously small quantities of feed in order to maintain slim profiles. For mature Runners, during the molt and for 6 to 8 weeks prior to a show, follow the feeding regime recommended above for Bali. To ensure plenty of exercise, our grassy pens are long, with waterers and feeders located at opposite ends. Runners fed unlimited quantities of a fattening diet, such as corn, meat bird grower, broiler grower, or cat food often become obese and do not show well.

To avoid kinked necks, Runners must be penned in cages that allow them to stand upright without bumping their heads on the top. The minimum height of show cages used for Runners is 27 inches. The carrying boxes we transport our Runners in are 32 inches tall to allow room for an abundant layer of bedding in the bottom.

Runners are the tightest feathered of all ducks, and it is easy for their wing feathers to become disheveled during transportation or as they are placed into show cages. Prior to judging, make sure their flight feathers are folded properly.

Comments

The graceful Runner is one of the most entertaining and useful members of the duck clan. When she sees a group of our Black Runners marching in a line, a friend of ours likes to declare, "Look, the miniature monks are going to town."

Runners are the most active foragers of all breeds, and will cover the largest territory in their hunt for snails, slugs, insects, and other edibles. Their active disposition is evident right from the start. When taking newly hatched ducklings from the incubator, one must move slowly and talk to them quietly to keep them from jumping overboard in their enthusiasm to explore their expanding world.

Runners make elegant show birds, entertaining pets, wonderful pest controllers, and are fine layers. Herding-dog trainers often start their young trainees on Runners. Without a doubt, Runners deserve their wide popularity.

SIX

MEDIUMWEIGHT BREEDS

Ducks in the Mediumweight Class typically weigh 6 to 8 pounds. They are general-purpose ducks, being fair to good layers, moderately fast growers, and producers of high-quality meat that is more flavorful and less fatty than that of most Pekins. When raised as broiler ducklings, they typically are feathered out under the wings and ready for butchering 2 to 3 weeks later than the fastest-maturing strains of Pekins. These breeds are well suited for situations where they can forage for some of their own food, and are then butchered once they have their full adult plumage at 16 to 20 weeks of age.

Mediumweight breeds are good foragers that are capable of eating large "banana" slugs. For many situations, they make excellent yard or pond ducks since they tend to stay close to home, do not fly under normal conditions, and are large enough that they are less likely than Bantam breeds to be preyed upon by winged predators. Typically, they have moderately calm temperaments and make fine pets. They are sometimes the first ducks used to train novice herding dogs.

In single male matings, a vigorous 1- to 2-year-old drake will normally successfully fertilize three to five ducks. In flock matings, four to six ducks per drake is typical. Many ducks of these breeds will get broody and hatch eggs if given a sheltered nesting area that is protected from predators.

Ancona

Most poultry raisers associate Ancona with the breed of chicken by that name. However, during the early part of the twentieth century, a breed of duck was developed in Great Britain and also called Ancona. They were likely developed from the same foundation stock as the Magpie and the genetic makeup of their plumage pattern supports this hypothesis. Some Anconas look identical to the old Belgium Huttegem duck.

Ancona ducks have been raised in the United States for several decades, and were exhibited in 1983 in Oregon. Although still rare, their numbers have been increasing since 1984 when they first became available to the public.

Description

This breed shares many physical characteristics with its apparent close relative, the Magpie. The main differences are that Anconas are an average of 1 to 1½ pounds heavier and have a bit stockier conformation. The Ancona has a medium-sized oval head; medium-length bill that is slightly concave along the top line; average-length and -diameter neck that is only slightly arched forward; and medium-length body carriage of 20 to 30 degrees above horizontal when relaxed.

The broken plumage pattern of Anconas is unique among ducks. Like Pinto horses and Holstein cattle, there is no set design. Any combination of white and color is acceptable as long as there are obvious broken areas on the head, back, sides, and underbody. The neck is normally solid white. At sexual maturity the bill should be spotted with green, often darkening to solid color with age. The legs and feet are orange with black or brown markings that increase with age.

Varieties

The Ancona varieties are: Black and White, Blue and White, Chocolate and White, Silver and White, Lavender and White, and Tricolored. The latter is any bird with white plus two or more colors in the plumage.

Selecting Breeders

As with all rare breeds, it is especially important to choose stock birds that are vigorous, free of physical deformities, and that have classic breed traits. Since the Ancona is an excellent laying breed, productivity should be given high priority in breeders. To produce the highest percentage of off-spring with the unique broken pattern, select birds with definite colored areas under their eyes and at least a bit of color on their chests. Avoid specimens that are either solid white or primarily colored with a white bib.

Selecting and Preparing Show Birds

Because the most distinctive feature of the Ancona is its broken-plumage pattern, choose show birds that have the boldest patchwork pattern. Avoid birds that resemble Magpies or Swedish. Cage-training prior to their first exhibition can help them show to their best advantage.

Comments

The patchwork-marked day-olds are adorable, and often are the first ones chosen when people are selecting from a group of assorted ducklings. Because their plumage designs are so diverse, seldom are two Anconas identical, making it easy to distinguish individuals in a flock.

Anconas are the best foragers and the most prolific layers of the Mediumweight breeds. Egg shell color can be white, tinted, blue, green, or spotted. They have proven to be extremely hardy and adapt to many environments.

An old trio of Black and White Anconas bred on our farm showing the body type and haphazardly broken plumage pattern typical of this breed.

Cayuga

In the heart of the Finger Lakes region of west central New York lies a slender, 40-mile-long glacial lake named after the Native Cayuga people. By 1863, the big black ducks being raised in the area surrounding the lake were called Cayugas. However, the seed stock for these ducks apparently came from southeastern New York, near the border of Connecticut.

A traditional account of the origin of the Cayuga says that a miller living in Dutchess County, New York, caught a pair of dark-colored ducks on his mill pond in 1809. Descendants of this pair were taken to Orange County, New York, where they multiplied. In approximately 1840, John S. Clark introduced them to Cayuga County where they thrived and gained a reputation as quiet, prolific layers that produced large meat birds.

In 1874, Cayugas were included in the first *American Standard of Perfection*. They were raised in significant numbers as general farm ducks, but by the 1890s were largely replaced by Pekins for the market duckling trade in the big cities. Today, Cayugas are raised in modest numbers for pets, decoration, meat, and eggs. The best exhibition strains are of superb quality and frequently win top honors. At large shows, Cayugas are often the most numerous of the Mediumweight breeds.

Many writers have suggested that the Cayuga is a descendant of the wild American Black Duck (*Anas rubripes*), which inhabits much of the eastern portion of North America. There is no hard evidence to support this theory, and much to disprove it.

American Black Ducks are black in name only; their plumage is dark brown with each feather edged with light brown. The true black plumage of the Cayuga is caused by a black mutation that is relatively common in Mallard derivatives. Furthermore, *Anas rubripes* drakes do not have curled sex feathers in their tails, whereas Mallards and their descendants (including Cayugas) have this diagnostic marker.

This Black Cayuga old drake that we bred won Grand Champion Duck at the 1991 APA National.

Description

Cayugas have typical duck conformation. Their bodies are moderately long, medium-wide, with good depth carried from the well-rounded chest to the moderately full abdomen. They have necks that are medium in length and diameter and only slightly arched forward. The medium-sized head is oval and the moderately long bill is slightly concave along its top line. Body carriage is approximately 20 degrees above horizontal. The crowning glory of the Cayuga is its resplendent green-black plumage.

Varieties

Black with green iridescence is the traditional color. A solid Blue Cayuga was developed and first shown in Oregon in 1984.

Selecting Breeders

Choose birds that are vigorous, strong-legged, and productive. Because many Cayugas are undersized, breeders that have good size, wide backs, and moderately long bodies with thick chests are highly valued. The bill, head, and neck should be in balance with a medium-sized duck. Some Cayugas have a row of feathers running down the outside of their legs. There are people who will not breed from these ducks; others will, if they are exceptional specimens otherwise.

Selecting and Preparing Show Birds

Look for ducks that have solidly built bodies and good size. Everything else being equal, the greener the plumage and the blacker the bill, legs, and feet, the better. (Since long exposure to bright sunlight can cause the plumage to develop purple reflections, black show ducks benefit from ample shade.) For maximum color, during the molt and for 6 to 8 weeks prior to a show, a good ration can be made by mixing 4 parts gamebird flight conditioner, 2 parts oats, 2 parts cat kibbles, and ½ part black oil sunflower seeds.

It pays to go over a Cayuga with a tweezers to remove any white feathers prior to taking them to a show. Pay special attention to the small feathers of the head and neck. One white feather in a black duck can make the difference between a bird winning first and not placing. (It is not permissible to remove large wing or tail feathers.)

Comments

For butchering, Cayugas must be plucked when they are in full feather to produce a clean-looking carcass. Pinfeathers can be eliminated altogether by skinning rather than plucking. In keeping with their ebony plumage, some Cayuga ducks produce their first eggs of the season encased in a black or gray cuticle. As the season progresses, egg color normally fades to light gray, blue, green, or even white.

Crested

Ducks adorned with feathery bonnets were depicted in art dating back more than 2,000 years. Some of the old Dutch painters included crested ducks of various colors in their rural scenes.

The crest is caused by a dominant mutation that has shown up with relative frequency throughout recorded history. Typically in any breed of Mallard descent, one duckling with a crest will hatch out of every 100,000 to 1,000,000 eggs hatched.

Today in some European countries, crested versions of most breeds are bred and shown. In North America, the White Crested was included in the first *Standard of Perfection* in 1874, with blacks being added in 1977. Even though Crested ducks are reasonably good layers and produce plump roasting birds, the majority are currently raised for pets and decoration. In many parts of North America, relatively few Cresteds are seen at poultry shows; due to their regal appearance, however, they always create considerable interest when they are exhibited. Demand for ducklings and breeding stock often exceeds supply.

Description

In all regards except for the crest, this is a normal-looking duck of medium size. The American Poultry Association Standard states that the carriage is "nearly horizontal," but almost every Crested duck I have raised or seen has stood at 20 degrees or more above horizontal. The size, shape, and placement of the crest in day-old ducklings is an accurate guide to what the head adornment will be on the adult duck.

Varieties

The two officially recognized varieties are the White and Black. However, ducks with crests can be found in most colors or patterns.

Selecting Breeders

First consideration should be given to birds that display abundant vigor, walk normally, and have no physical deformities such as severely kinked necks, roached backs, or wry tails. Moderately long, plump bodies with deep chests are desirable.

It is not necessary for breeders to have large crests for them to produce some ducklings with large headgear. Therefore, it is more important to select breeding birds that have firmly attached crests with smooth, rounded surfaces and no obvious divisions, that are centered as high on the skull as possible. No matter how good the parents are, some ducklings will be produced that have small, misshapen, lopsided, loosely attached, or totally absent crests.

Selecting and Preparing Show Birds

Everything else being equal, the larger, rounder, and more firmly attached the crest, the more valuable the bird. Look for crests that are attached fairly high on the back of the skull. Ideally, the front edge of the crest is even with the eyes. Most judges prefer full-sized birds (7 pounds for old drakes) with solid build, but excessive size and coarseness resembling Pekins should be avoided.

To maintain the crest's good condition during the show season, birds should be kept in a clean environment and the females (if necessary) separated from overly active drakes that may pull out feathers during mating. During the molt and for 6 to 8 weeks prior to a show, a good ration can be made by mixing 4 parts gamebird flight conditioner, 2 parts oats, and 2 parts cat kibbles. Check crests for lice or mites and treat as needed. Show cages used for Crested ducks must be a minimum of 27 inches tall so that the crest does not rub against the top of the pen.

Comments

Many people are attracted to the royal appearance of Crested ducks. However, these birds do have some inherent challenges. Crests in ducks (unlike geese and chickens) are caused by an incompletely dominant autosomal gene, which has been given the symbol Cr. When two ducks with crests (Cr/cr) are mated together, approximately one quarter of their embryos (those homozygotes with two genes for crest, Cr/Cr) die. Typically, of those that survive, approximately two-thirds will have crests (Cr/cr) and one-third will be plain-headed (cr/cr). In those birds that carry the gene for crest (Cr/cr), their head adornment may range from being invisible to measuring 5 or more inches in diameter. Furthermore, ducks that carry the gene for crest are susceptible to skeletal deformities (kinked necks, shortened bodies, roached backs, wry tails) and balance problems. The balance difficulties may appear immediately at hatching time or not until hours, days, weeks, or months later.

Fortunately, the majority of Crested ducks are normal and have a life expectancy similar to that of other breeds. Egg shell color can be white, tinted, blue, or green.

A good breeding trio of White Crested ducks, bred by the late Henry K. Miller. Breeders do not necessarily have to have show-type crests to produce some offspring with good crests.

Orpington

Englishman William Cook, the famous poultry breeder and promoter who lived near Orpington in the southeast county of Kent, originated the Orpington breeds of both chicken and duck. He created his duck breed by blending together the Cayuga, Runner, and production-bred Aylesbury and Rouen, and then carefully selecting those progeny with the desired characteristics. At the turn of the twentieth century, buff-colored plumage was highly prized in the British poultry world, and Buff Orpington ducks were developed to cash in on this fad. In England, they were first shown in 1907.

In quick succession, the Blue Orpington was introduced in 1908, followed by the Black and Chocolate. (The last three varieties had white bibs on their chests.) In England, where they were promoted by the Orpington Duck Club, this breed performed well in the popular egg laying contests (at the Harper Adams Agricultural College Laying Tests of 1928, Buff Orpingtons averaged 204 eggs in 12 months) and were raised in fair numbers as quick-maturing meat ducks.

Orpingtons made their appearance in the United States in 1908, when their originator showed a pair of Buffs at the prestigious Madison Square Garden Show in New York City. In 1914, they were admitted to the *American Standard of Perfection* under the name of Buff. This is an odd and confusing name that breaks Standard nomenclature procedure for poultry. (In no other instance is a color used as a breed name.) Both for clarity and historical perspective, Orpington should be used to designate the breed, and Buff the variety. Blue Orpingtons were advertised and shown occasionally in North America during the first half of the twentieth century, but apparently were ultimately absorbed by the Swedish breed.

No one ever questioned the fine practical qualities of the Orpington, but they arrived in North America 35 years after the Pekin and were never able to compete with them as a commercial market duck. Currently, Buff Orpingtons are raised in modest numbers as an all-purpose duck, and at shows they are typically numerically second to Cayugas in the Mediumweight Class.

Description

In conformation, the Orpington is similar to the Cayuga. The only significant differences between the two breeds are that the Orpington should have a slightly longer body and its bill should be attached slightly higher on

A pair of Buff Orpingtons, bred by the late Henry K. Miller, with excellent type and even color.

the head with a nearly straight top line, reminiscent of its Runner heritage. Body carriage is approximately 20 degrees above horizontal.

Varieties

In North America, Buff is the only color currently known to exist.

Selecting Breeders

Many Orpingtons are significantly under Standard weight and have bills with excessively concave top lines. Therefore, full-sized birds with straight bills that are attached high on the head make valuable breeders. When possible, avoid birds with obvious white on their necks or elsewhere on the surface of their plumage.

Selecting and Preparing Show Birds

The bird with the best plumage color often wins. Look for a medium shade of warm fawn-buff that is as uniform as possible in all sections and has a minimum of blue or gray shading. A white neck ring is a disqualification. Top show prospects also have good size, long bodies, deep chests that are nicely rounded, and nearly straight bills that are attached fairly high on the head.

During the molt and for 6 to 8 weeks prior to a show, you can use the same feeding recommendations as outlined for Crested to encourage excellent condition. Because buff-colored plumage is susceptible to fading, it is a common practice to keep ducks of this color out of midday sun for 6 to 8 weeks prior to and during the show season.

Comments

A group of well-bred Buff Orpingtons is a lovely sight, especially when they are foraging on a carpet of green grass or loafing on a blanket of snow. They do not show soiling as easily as white-plumaged ducks, and yet their pinfeathers are not unsightly when they are dressed for meat. Egg shell color is usually white or tinted. When out of a high-producing strain, Orpingtons can provide a good supply of eggs, medium-sized roasting ducks, and beautiful show birds.

Swedish

Blue-colored ducks have been known in Europe for centuries. Tradition held that blue ducks were exceptionally hardy, superior meat producers and difficult for predators to see. As early as 1835, the foundation stock of the Swedish duck was being raised by farmers in Pomerania, which at that time was part of the Swedish Kingdom, but today straddles northeast Germany and northwest Poland.

Swedish ducks were imported into North America in 1884 and included in the *American Standard of Perfection* in 1904. During the intervening years, they have been raised in modest numbers as general-purpose farm ducks and for pets, decoration, and exhibition.

The original written *Standard* called for the two outermost flight feathers on both wings to be white, with the remainder of the wing being blue. This unreasonable specification discouraged many people from showing this fine breed. In 1998, the Standard was changed to allow either two or three of the outer flights to be white. This revision makes it slightly more likely to meet the Standard specifications for wing color. However, both breeders and judges need to keep in mind that the color and pattern of the Swedish is one of the most challenging to perfect, and that placing too much emphasis on such a minor detail as the exact number of white flights is a serious detriment to the breed.

Description

Compared with Cayugas and Orpingtons, Swedish are broader and not quite as long in the body, giving them a slightly stockier appearance. However, short bodies are not desirable in this breed, although it is a commonly seen fault. The head is oval-shaped and medium to medium-large in size. The medium-length bill should be nearly straight along its top line, as in the Orpington. But in the Swedish, the bill should be attached to the head slightly lower, with a bit more rise to the forehead. Body carriage is approximately 20 degrees above horizontal.

Varieties

Blue is the only Standardized color, but Swedish also produce Black, Silver, and Splashed offspring.

A Blue Swedish young duck with typical type, bred by Perrydale Acres of Oregon. Ducks usually have smaller white bibs than drakes.

Selecting Breeders

Top priority should be given to vigorous, strong-legged birds with solid, well-muscled bodies of good size. Common faults to avoid include short, narrow, or shallow bodies, tails that are constantly cocked upward, narrow snakey heads, and bills that are excessively long or have a distinctly concave top line.

Selecting and Preparing Show Birds

Judges admire a Swedish with good size, adequate length, rich blue plumage with as little foreign color as possible (even the best specimens have a bit of black mixed with the blue), and a well-defined white bib.

During the molt and for 6 to 8 weeks prior to a show, you can use the same feed mix recommended for Crested. Although blue plumage is more stable than buff, over time sunlight will discolor it. Therefore, keeping blue ducks in shade during the intense sunlight hours of midday for 6 to 8 weeks prior to showing will improve feather quality and clarity of color.

Comments

When well bred, Swedish are a beautiful duck that rewards the person who has the patience to perfect both the color and pattern of this hardy breed. Egg shell color can be white, tinted, blue, or green.

SEVEN

HEAVYWEIGHT BREEDS

The older breeds in the Heavyweight Class traditionally were raised primarily for meat. Typical weights range from 7 to 12 pounds, although some individuals will tip the scales at considerably more. Characteristics that most of these breeds share include heavily muscled bodies, exceptionally fast growth, moderate rate of lay, the tendency to stay close to home if well fed, and calm temperaments.

Ducklings will reach their full size only if they are raised in the right environment and supplied an ample quantity of feed that meets all of their nutritional requirements. Inadequate nutrition can result in birds reaching as little as 60 percent of their genetic weight potential.

On the other hand, ducks that are going to be used for breeding stock or pets will live longer, reproduce better, and have fewer leg problems if they are not pushed for the fastest possible growth rate or maximum body weight.

Over time, there is a tendency for large breeds to gradually lose size. Therefore, it is important to select breeding birds in each generation with good size; however, the very largest drakes are sometimes not fertile. A common practice is to select for breeding purposes the largest worthy females, mating them with active drakes that have good size, but are not necessarily the biggest. When selecting potential breeders, be aware that occasionally there are birds that appear to be huge females, but in fact are androgynous and will never lay eggs.

For good production, it is important that Heavyweight ducks are in good flesh but not obese as they enter the breeding season. Obesity can cause lameness, infertility, and prolapsed oviducts. In breeds that have profusely feathered abdomens (primarily Pekin and deep-keeled Rouen and Aylesbury), fertility is often improved if the feathers 3 inches on all sides of the vent are trimmed short with scissors.

In single male matings, a vigorous 1- or 2-year-old drake that is not overweight and is consuming a breeder diet that encourages good fertility normally will service three to four ducks. (In the very large, deep-keeled, exhibition-type Rouen and Aylesbury, pair or trio matings are usually the most successful.) In flock matings, four to five ducks per drake normally gives good results. Some ducks in this class are good to excellent natural mothers if allowed to nest in a safe location. Extremely heavy ducks may get broody, but they have a tendency to crush their eggs or ducklings.

Appleyard

The Silver Appleyard was developed by Reginald Appleyard in the 1930s at his famous Priory Waterfowl Farm near Bury St. Edmund, in West Suffolk County, England. In a farm brochure from the 1940s, Mr. Appleyard states that his purpose was to "make a beautiful breed of duck, with a combination of beauty, size, lots of big white eggs, and deep, long, wide breast."

Following World War II, interest in poultry breeds declined and the Appleyard almost disappeared. Tom Bartlett of Folly Farm near Burton-on-the-Water, England, was instrumental in reviving this fine breed during the last third of the twentieth century.

Appleyards were brought to the United States in the late 1960s. However, it was not until 1984 that they were made available to the public and exhibited in Oregon. By 1988, Appleyards had gained sufficient popularity that at the Boston Poultry Show they constituted one of the larger classes of ducks. The qualifying meet for admission into the American Poultry Association (APA) *Standard of Perfection* was held October 1998 at the Minnesota State Poultry Association Show in Hutchinson, with 106 Appleyards shown by 12 exhibitors. Currently, Appleyards are being raised in increasing numbers for exhibition, pets, decoration, eggs, and gourmet roasting ducks.

Description

The Appleyard is a moderately large and sturdily built duck that should not have the extreme coarseness of features nor the loose feathering of the Pekin. The bill is of medium length with a nearly straight top line. Nearly

This pair of Silver Appleyards bred on our farm display typical type and color. Winners of Best and Reserve of Breed at the 1999 Western Waterfowl Expo.

oval with a slight forehead crown, the moderately large head is attached smoothly to a neck that is moderately thick, medium in length, and slightly arched forward. The compact body, which is carried at 15 to 25 degrees above horizontal, is broad and deep with a prominent chest that is smooth and well rounded.

In Great Britain, Appleyards are bred and shown both plain-headed and with crests. Even though we have only bred from plain-headed birds for many generations, crested ducklings are still produced with fair regularity from some matings.

Varieties

Silver is the only color variety, but Mr. Appleyard created both a large (described here) and a miniature version (which weighs only 1¼ to 1¾ pounds).

Selecting Breeders

Appleyards are the most active foragers and the best layers among the Heavyweight ducks, so it is important to use breeders that are robust, strong-legged, and excellent producers of large eggs. Many Appleyards are under-sized, so birds with big, solid-muscled bodies are valuable. On the other hand, excessively large size (which currently is a rare problem) hampers foraging and laying ability, and is not desired.

Selecting and Preparing Show Birds

Big, solid birds with no keel on the breast, smooth silky plumage, and proper color will catch the judge's eye. If provided a clean environment, proper diet, and clean bathing water, Appleyards will usually get themselves in good condition for showing. During the molt and for 6 to 8 weeks prior to a show, feeding a ration consisting of 4 parts gamebird flight conditioner and 1 part cat kibbles will encourage excellent feather condition.

Comments

Compared with Pekins, Appleyards do not grow quite as fast, but they have much more interesting colors, are better layers, produce meat with more flavor and less fat, are better foragers, and are more likely to incubate and hatch their eggs. Most females lay white-shelled eggs. Appleyards are one of the best all-purpose breeds of ducks and adapt to a wide range of environments.

Aylesbury

In the Vale of Aylesbury, less than 40 miles northwest of London, Buckinghamshire "duckers" were renowned by 1800 for the production of England's finest white-skinned market ducklings. The famous pink-billed ducks of this region were originally known as White English, but by 1815 they were being called Aylesbury. Along with their unusual bill and skin color, these ducks were held in high esteem for their large size, ability to produce eggs in the winter months, fast growth, light bone, and a high percentage of mild-flavored meat when butchered at 6 to 9 weeks of age.

For years it was thought that this unique duck could be raised successfully only in its home region where there was an ample supply of a pumice-

An 8-month-old Aylesbury duck with great body depth and excellent back profile (top line). Bred by Good Shepherd Ranch in Kansas.

like white gravel found in the local streams. The common belief was that by digging their bills into this peculiar grit and consuming some of it, the birds' bills and flesh were lightened to the pale pink color so highly prized by the British epicures. Over time, it was discovered that the Aylesbury could be raised in other locations, although not always with the same degree of pale color in the bill and flesh.

As poultry breeders began showing their prized fowl, some strains of Aylesbury were selected for greater size and deeper keels. Eventually two types evolved — production-bred market birds and standard-bred show specimens. At the first British poultry show held in 1845, two classes were provided for ducks: "Aylesbury or other white variety" and "Any other variety."

The Aylesbury was one of the first duck breeds to arrive in the United States from Europe. However, because yellow-skinned market birds have traditionally sold for higher prices in North America, the commercial use of Aylesbury never flourished in Canada or the United States. Furthermore, nineteenth-century American duck growers considered the Aylesbury to be less hardy than other breeds. In retrospect, this perceived lack of hardiness was likely caused by diets that were deficient in vitamins A and D_3.

Aylesbury were exhibited at the inaugural poultry show in America, held in 1849 in Boston, Massachusetts. In 1874, they were included in the first *American Standard of Perfection*. Although not exhibited in large numbers at most shows, Aylesbury have had a devoted following over the decades among specialty breeders and continue to win their share of honors.

Description

With a long body, horizontal carriage, and pendulous keel that runs from stem to stern and brushes the ground in top specimens, the Aylesbury is a bona fide low rider. Ideally, the top line of the back is only slightly convex from the shoulders to the tail. However, many Aylesbury, especially those that are extra large and long-bodied, have a fair amount of arch to their backs when they are standing still and relaxed. The unique head has been compared to that of the American Woodcock, with a flattened forehead, high-set eyes, and a long, straight bill that ideally measures 3 to 3½ inches or more in length. The neck is moderately long with a slight forward arch when the bird is relaxed.

The plumage is the whitest of all ducks, the feet and legs orange, and the eyes dark grayish-blue. The natural color of the bill is pink in day-old ducklings and pale yellow to blanched yellowish-pink in nonlaying adults. After they have been laying for a time, bill color in most females will fade to very pale pink, usually with black streaks or spots in the bean.

Varieties

Aylesbury are bred in White only.

Selecting Breeders

The Aylesbury is a complex breed that is a far departure in size and conformation from the wild Mallard. It is one of the more challenging breeds to perfect. Ideal breeders will seldom — if ever — be found, so putting together a mating is a balancing act. The rule of thumb is: Try to avoid mating together two birds with the same obvious faults.

It is critical that breeders have strong legs, display good vigor, and are free of genetic physical deformities. Great length and depth of body are desired, along with good width of back. However, the largest males with the most exaggerated type are sometimes not fertile, so it can be risky relying on them as breeders.

Common faults to watch for include: small size; bodies that are short or shallow; high-crowned foreheads; bills that are short and/or distinctly concave along the top line; body carriage that is obviously elevated in front; wings that slide far down the sides of the body or are constantly held excessively high with the tips crossed over the back; and keels that are shallow, short, wry, uneven along the bottom line, or do not have a prominently curved bow on the front.

When possible, avoid using drakes that have black pigment anywhere on their bills, especially if they are 2 years or younger. (Keep in mind that injuries to the bill sometimes cause transitory discoloration.) Be aware that if excessive effort is made to eliminate all black from the bills of sexually mature females, the productivity of the strain will usually decline.

Selecting and Preparing Show Birds

A good show specimen has a profile resembling a rectangular cement block with rounded edges, a long, straight, whitish-pink bill, and satin-white plumage. To obtain the ideal bill and plumage color, Aylesbury need to be kept for at least 6 to 8 weeks prior to showing in a clean environment with nonstaining bedding, provided clean bathing water, kept out of sunlight, and fed a diet that is low in yellow pigments. Feed ingredients that help promote pure white plumage and pale-pink bills include oats, barley, white wheat, white rice, roasted soybeans or soybean meal, fish meal, meat meal, dried milk, and distillers' solubles. A ration that promotes good health and encourages decent feather and bill color consists of 4 parts gamebird flight conditioner, 2 parts oats, 2 parts white wheat, and 2 parts fish-based cat kibbles.

For ducks that are kept in shade and consume a diet low in yellow-pigmented ingredients, it is critical that a good multivitamin poultry supplement be given in order for birds to maintain good health and physical appearance. The amount of vitamin supplement required depends on the vitamin levels in the regular diet. In general, if the vitamin supplement is given in an amount that provides a daily dosage of 100 to 200 ICUs of vitamin D_3 and approximately 1,000 ICUs of vitamin A per bird, the concentration of the other vitamins in the supplement will be appropriate.

Comments

The unique bill, skin, and plumage color of the Aylesbury results from the interaction of genetics, diet, and environment. Aylesbury carry a dominant autosomal gene that has been given the symbol Y. This gene reduces a bird's ability to deposit yellow pigment in its skin and feathers. Newly hatched ducklings with the Y gene have pink bills.

Ducks carrying the Y gene normally develop yellow-tinted plumage and bills if they are exposed to direct sunlight, bathe in hard water, are raised on soils that are red and/or have elevated iron levels, or consume feeds that are high in yellow and orange pigments. White-skinned ducks have elevated requirements for vitamins A and D_3. However, under many circumstances, supplemental feeding of these vitamins is required only when birds have limited access to direct sunlight or consume diets low in natural yellow pigments.

Fertility is usually improved if swimming water at least 6 inches deep is available. Females typically lay white, tinted, or green eggs. Aylesbury are easily tamed and make responsive and unique pets.

Muscovy

The wild Muscovy (*Cairina moschata*) is a nonmigratory native to Mexico, Central America, and much of South America east of the Andes Mountains. Being a tree duck, they roost in trees at night and spend hours preening on elevated perches after their daily bath. Preferred nesting sites are cavities in tree trunks and branches.

One of the larger wild ducks, drakes average 4¼ to 6 pounds and ducks 2¼ to 3 pounds. Both sexes have blackish plumage with metallic green sheen. Juveniles especially have dark brown or bronze on the chest and underbody. The brilliant white wing badges develop gradually with age, being the most visible during preening and in flight, when they flash unmistakably. The area around the eyes of the sexually mature drake is covered with smooth, black skin that is jeweled with one or more red spots. The junction of the forehead and bill is adorned with a black, fleshy nub that is highlighted with bright red. Females are nearly devoid of facial skin patches with only a rudimentary nub on the base of the bill. In both sexes, the feet and legs are dusky black. The bill is black, with a bluish-white or pinkish-white band across the end, and then tipped with a blackish bean.

This Blue Muscovy old duck possesses the dark lacing desired on the body plumage. Bred by Barry Hampton, and owned by Gary and Kari Bennett.

By the time Columbus arrived in the New World, Muscovies had most likely been raised in domestication for centuries. In 1514, Spanish Conquistadors found the native people raising several colors of Muscovies on the northern coast of Colombia. From their home in the Americas, Muscovies probably traveled to Africa and then to Europe, where they arrived by at least the mid-1500s.

A host of theories have been advanced as to how this native American became known by a Russian name. A reasonable suggestion that is logistically possible maintains that these birds were transported to England by the Muscovite Company during the 1500s. Since it was a common practice to attach the importer's name to products it traded, it would have been an easy transition from Muscovite Duck to Muscovy.

In North America, the Muscovy has a long history, especially in Mexico, of being raised as a self-reliant duck that hatches and often raises multiple broods a year with a minimum of assistance from humans. However, they did not become a major market duck in Canada or the United States until the later part of the twentieth century. Today, as the demand for leaner meat and liver paté steadily increases, the breeding and marketing of Muscovies and their hybrids are expanding.

When the first American poultry show was staged in Boston in 1849, three people exhibited Muscovies. In 1874, when the first *American Standard of Perfection* was compiled, the White Muscovy was included, even though colored birds were more common. Exhibition Muscovies have experienced a renaissance in both quantity and quality during the past decade.

Description

It only takes a glance to realize that domestic Muscovies truly are a unique breed. A closer look reveals the physical characteristics that allow its wild ancestors to be at home in tropical rainforests.

The medium-length bill is relatively narrow, with large nostrils, a fairly concave top line, and an extra large nail (bean) at the tip. The head of the mature drake is massive and its face covered with red skin. The duck has a medium-sized head and smaller facial skin patches. Both sexes can raise the feathers on top of their heads into a crest when excited, flirting, or angry. The male's crest feathers are longer and stylishly waved.

Compared with other breeds, the body is flattened, heavily muscled, and extra wide across the shoulders. The wings are very wide and moderately long, with the tips being more rounded than in other breeds. The tail is long and broad. (The wings and tail were designed for flying through heavily forested terrain.) The long toes are webbed, amazingly strong, and tipped with talon-like claws.

The size differential between the sexes is remarkable. At hatching, Muscovy ducklings are the same size. By 3 weeks of age, females weigh approximately 65 percent as much as the males, at 12 weeks 55 percent, and at 26 weeks 45 to 50 percent. For comparison, in other domestic breeds and wild species that nest on the ground, ducks typically weigh only 8 to 12 percent less than drakes at maturity. In species that nest in tree cavities, such as American Wood Duck, Bufflehead, and Common Merganser, females on average weigh 23 to 31 percent less than males. It is possible that in tree-nesting species, smaller females are more likely to find safe nesting cavities.

As Muscovies go about their daily activities, they frequently wag their tails from side to side and thrust their heads front to back. A less frequent but still common display includes pointing the bill to the sky and rapidly moving the mandibles as if catching invisible raindrops.

When Muscovies are on uncrowded free range, there is usually a minimum of serious fighting. However, they can battle with amazing fervor if their territory is too small. Sometimes they face each other like fighting chickens, leaping into the air and striking at one another with their claws. Other times, they fight by pecking each other with their bills and pounding with the bony protuberance on the leading edge of their wings. When breeding pens of Muscovies are adjacent to one another, it is usually beneficial to have solid metal or wooden dividers at least 2 feet high along the entire length of the pens so that the drakes in adjoining pens cannot see each other and attempt to fight through the fence.

Muscovy ducklings can be identified from other breeds by their medium-sized bill with a particularly large bean on the tip, flattened bodies, distinctive trilling vocalization, larger and stiffer tails, and needle-sharp claws. (When naturally hatched in tree cavities, the ducklings use their stiff tails and sharp claws for climbing out.)

A Memorable Moment

To be on that river deep in the interior of Guatemala was to experience a real-life fantasy — miles of meandering milky green water, dozens of palm-sized, electric-blue butterflies, and a towering ceiba tree with a pair of toucans hopping along a massive branch. As our canoe rounded the bend, the sleek, black duck vaulted from its perch 15 feet up in the snag jutting out of the river and flew straight into the jungle. Our first sighting of a wild Muscovy! From that brief encounter, it was concluded that we had just crossed paths with a wily, old bird. The first hint of its veteran status was the bold white forewing patches that only older birds display. Second, rather than taking the open flyway down the river, which would have made it an easy target for the many hunters who plied these waters, it vanished straight into the woodland. After having seen domestic Muscovies in nearly every village and marketplace we had visited, it was a thrill to finally glimpse one of the untamed originals.

Varieties

White, Black, Blue, and Chocolate are recognized by the APA. Wild Type, Silver, Lavender (sometimes called Self Blue), Buff, Blue Fawn, Lilac, Pastel, Barred, Rippled, White-headed, Magpie, Splashed, and others are also raised.

Selecting Breeders

In order to choose breeding stock wisely, you need to clearly have in mind your main purpose for raising Muscovies. Remember, there are various distinct strains, and characteristics such as size, growth rate, feed efficiency, egg production, maternal instincts, and foraging ability can vary drastically among them.

If your goal is to have self-reliant birds that will find much of their own food and hatch and raise their own babies, the main criteria you should consider in making your selection should include vigor, foraging ability, and maternal instincts. Birds that are medium- to dark-colored have better camouflage in most settings. Medium-sized birds that can fly are typically better foragers and are also less likely to be caught by predators. Many commercial strains have had the maternal instincts bred out, making them worthless as natural mothers.

If your goal is to produce the maximum number of market birds, then the primary selection criteria should include good egg production, fast growth, efficient feed conversion, heavy breast muscling, and a high meat-to-fat ratio. For many markets, birds with good size are desirable. However, as size increases beyond a reasonable level, it is normal for productivity, feed efficiency, and growth rate to decrease. Huge show birds usually do not make efficient producers of market ducks.

For the production of show birds, the main characteristics that you should select for include massive size, good type, proper plumage color, and well-caruncled faces (caruncles are the bumpy skin on the face) with a minimum of black. If the Muscovies are descendants of a strain with good faces, it is not necessary for breeding females to have exceptionally large facial skin patches to produce offspring with good exhibition faces. In fact, it is not unusual for females with male-like caruncling to be sterile due to a hormonal imbalance.

Selecting and Preparing Show Birds

Good show specimens are massive throughout, with long bodies and wide shoulders. The head is wide across the skull, and in mature birds the face is covered with well-defined caruncles that are grainy right up to the eyes. In old drakes, a fleshy knob the size of a grape or larger lies above the nostrils on the upper mandible. The knob on hormonally normal females is much smaller. The caruncling should be balanced on both sides of the head. It is fine for the caruncling to extend down the neck in old drakes, but it is a serious fault if it blots out the crest feathers on top of the head or makes it impossible for the bird to see.

To have Muscovies in top condition for showing, provide them with a well-drained, clean pen with mounds of clean straw for them to perch and preen on, clean bathing water (dirty water and mud can ruin a White Muscovy for showing), plenty of shade (although some direct sunlight is needed to get the brightest red color in the face), a daily feeding of succulent greens such as lettuce, chard, and tender grass clippings, and a ration consisting of 2 parts gamebird flight conditioner, 1 part oats, and 1 part cat kibbles, plus insoluble grit. A teaspoon per bird of canned cat food several times a week is greatly enjoyed once they develop a taste for it and seems to help improve feather quality. To show off to their best advantage, large drakes need to be penned in double coops at shows.

Comments

Muscovies are often described as mute. Although this is not literally accurate, they are by far the quietest of all breeds. Females occasionally quack weakly, but mostly use a variety of soft chirps and whistles to communicate. The drake's primary vocalizations are hoarse, breathy exhales of varying lengths.

Muscovies can appear to have extremely laid-back personalities. When walking through their territory, you can almost step on them before they move. But as soon as they know your intent is to catch them, they are uncannily quick and adept at avoiding your grasp. Because they do not herd as well as other ducks, a good way to catch them is to walk them into a V-shaped corner one at a time and then snatch them with your hands or a long-handled fishnet. Hand-tamed pets can often simply be picked up off the ground.

Muscovies are by far the strongest of all ducks, and with their powerful legs and long claws are capable of inflicting painful scratches on your hands and arms when they are caught. I wear long-sleeved denim coveralls and leather gloves when handling them.

Some people extol the Muscovy as being the sweetest-tempered of all ducks, while others insist that they are nothing short of the fabled Tasmanian Devil incarnated in duck form. Depending on the particular bird and the circumstances, both descriptions are reasonably accurate. Some drakes live amicably for years in yards with all manner of fowl and animals, whereas others will mercilessly drag down and attempt to mate with critters of many shapes and sizes. Full brothers can have personalities that are polar opposites. Some females, particularly when new birds are introduced to their territory, will attack like gamecocks. When serious assaults persist for more than a brief time, either the victim or the attacker should be removed. While I have found no failproof way to predict personality, older individuals and birds that have been raised in small groups tend to be the most aggressive. Conversely, those that have been raised around diverse creatures and in larger groups typically are more tolerant.

Most females, if not overly fat, are fair to good flyers and will perch on fences and buildings. If they have sufficient food and feel safe, they normally will not leave their home domain. The annual clipping of the ten primary flight feathers of one wing prevents sustained flight. Most mature drakes are too heavy to become airborne for more than a short distance. However, if sufficiently motivated, they are capable of climbing over 3- to 4-foot-high wire fences.

Muscovies enjoy swimming, and if permitted to do so will bathe daily. However, their feathers do lose water repellency easier than most ducks. Individuals with poor feather condition or those unaccustomed to swimming can become waterlogged if forced to stay on water.

Their tropical roots notwithstanding, Muscovies can tolerate considerable freezing temperatures as long as they are protected from cold winds and have a draft-free, well-bedded shelter to sleep in at night. In wet weather they should have access to a dry pen.

Muscovies are highly prized for their meat and are easier to pick than other ducks. They have exceptionally broad, well-muscled breasts and are one of the leanest of all waterfowl. We have found that their meat makes a fine "beef" stew, while cured and smoked Muscovy is similar to lean ham.

Muscovy eggs can be hatched in incubators, but for good success incubation temperature and humidity must be precise. In my experience, the best results are obtained when their eggs are incubated by themselves, the temperatures and humidity levels are recorded each day, and from the 15th to 32nd day of incubation the eggs are cooled once daily by spraying them with lukewarm water. At the end of the incubation period, the daily temperature and humidity readings are added up and then divided by the number of days. If the hatch was a success, the incubator should be operated at the same average daily temperature and humidity levels for the next batch of eggs. If the eggs hatched poorly, then the temperature and/or humidity should be adjusted.

The ducklings of this species have a reputation for being difficult to ship. However, we have had good success shipping them across the continent. The keys to shipping Muscovy hatchlings are proper incubation, mailing them no more than 12 to 24 hours after they emerge from the shell, and sending them only to destinations that they will reach in 48 hours or less.

Females are usually dedicated mothers, and it is not unusual to see them with 10 to 16 ducklings in tow. In fact, Muscovies are such reliable broodies that they are often used to hatch the eggs of other waterfowl, including rare species of wild ducks, geese, and swans. Their eggs are typically white to tan and have an incubation period of 33 to 35 days. Muscovy drakes normally fertilize more ducks than other Heavyweight males. In single male matings, a drake can usually fertilize five to seven ducks. In flock matings, six to ten ducks per drake usually gives good results.

As long as they are not overly inbred nor out of a commercial strain, Muscovies are the most self-reliant of all poultry species in many situations. They are active foragers, consume more grass than other ducks, and are first-rate fly catchers.

Pekin

The ancient people of China were among the first to domesticate the Mallard, probably before 1000 A.D., and developed ducks into an important food source. Over time, the Chinese became some of the world's most advanced duck raisers, developing sophisticated artificial incubation techniques and producing various types of ducks, including the Pekin.

The arrival of the Pekin in the United States is described in a letter written by James E. Palmer that was published in the September 1874 issue of

The Poultry World and quoted by John H. Robinson in his book *Ducks and Geese for Profit and Pleasure* (1924). In his account, Palmer states that he was asked by an American businessman, a Mr. McGrath, to transport 15 large white ducks from Shanghai to the United States. The ducks had been hatched from eggs acquired in Peking (today known as Beijing). Three drakes and six ducks survived the 124-day oceanic trip, landing in New York City in March 1873.

In 1874, the newly formed American Poultry Association published its first book of Standards, and included the Pekin. Within a short time after their introduction into North America, they became the primary breed raised for the production of market ducklings. Their popularity continues today, with Pekins being raised in greater numbers than all other breeds combined. Their commercial success is due to a combination of traits that include hardiness, large size, unsurpassed growth rate (up to 8.5 pounds in 7 weeks), high feed efficiency (2.6 pounds of feed per pound of body weight), high fertility, excellent hatchability, calm temperament, and white plumage.

As show birds, Pekins are popular for 4-H and Future Farmers of America projects, especially the market classes. In open class competition at large shows, they are common, but often not as numerous as Rouen or Muscovy.

These huge 7-month-old, standard-bred Pekins weigh 12 pounds (the drake) and 11 pounds (the duck). The drake won Reserve Champion Heavy Duck at the 1999 Northwest Winter Classic. Bred by Gene Bunting.

Description

From early illustrations and written accounts, it is clear that the average Pekin raised in North America today has not changed significantly in appearance. The classic American Pekin has a thickset body that is carried at a jaunty 35 to 45 degrees above horizontal. The underbody is wide, smooth, and keelless, with an ample abdomen that is profusely feathered. Unlike other breeds, the tail habitually sticks up above the line of the back. The head is large with full cheeks, and especially in females the forehead feathers rise abruptly from the base of the bill to form an elevated crown. The neck is thick, moderately long, and often appears (mostly due to the way the bill, head, and neck flow from one to the other) to be arched forward. The bill should be of medium length, thick enough to be in harmony with the massive bird, and nearly straight along the top line.

In England and Germany, where the Pekin was introduced in 1874, Pekins stand almost vertical, have short plump bodies, short thick necks, chubby heads, and short bills. German Pekins were imported into North America in the 1990s, and their offspring, both pure and crossed with American-strain birds, have been exhibited at some shows.

Varieties

Pekins are bred in creamy White only.

Selecting Breeders

When Pekins first arrived in the Americas, the main selection criteria were vigor, strong legs, large size, and fast growth rate. Later, females were housed individually in narrow runs and those that were the best layers were kept for breeding. Then interest grew in producing strains that were the most efficient at converting feed into body mass. While commercial breeders continue to select for all of these traits, there is a growing emphasis today on developing market Pekins with less fat and more breast meat. The result is that a number of distinct types of Pekins are currently available, each with unique market traits. When acquiring breeding stock, choose the strain(s) that best fit your needs. In general, the largest and fastest-growing strains have the most body fat, whereas those that are the leanest grow a bit slower and consume more feed per pound of bird.

Pekins that have been bred for exhibition are larger and more massive than commercial strains. In this breed, individuals that are good show birds normally also produce good exhibition offspring. Because they are large, always make sure to choose breeders that have stout legs and walk normally.

Selecting and Preparing Show Birds

Top show specimens stand at approximately 40 degrees above horizontal and are massive in all sections. They have long, thick bodies, barrel-shaped chests, and powerful heads and necks. (The exhibition Pekin has the largest head and neck of all breeds.) To add style and mass to the head and neck, it is desirable for the feathers of the back of the head and upper neck to project back or slightly upward, forming a mane.

The Pekin is the only white duck that should have a strong cream or yellow cast to its plumage. The creamy hue is caused by ingested yellow and orange pigments. Therefore, diets rich in green feeds, yellow corn, marigold petals, and other yellow-pigmented foodstuffs will increase a bird's depth of color, especially if consumed 6 to 8 weeks prior to and throughout the molt and regrowth of feathers. A good show ration to enhance color and body size consists of 4 parts gamebird flight conditioner, 3 parts yellow corn, 2 parts cat kibbles, and 1 part rabbit or alfalfa pellets. Yellow pigment in the plumage bleaches easily. The yellowest Pekins I have seen were exhibited by a poultry judge whose property had red clay soil.

Comments

Pekins are ideally suited for situations in which the quickest-maturing meat ducks are desired. They make good pets and are often kept on ponds "just for pretty." Females are talkative and lay white or tinted eggs.

Rouen

Among the tame ducks raised by French farmers several hundred years ago were some resembling large Mallards. Around 1800, these ducks reached England, where they were variously called Rhône (an area in southwest central France), Rohan (a Catholic Cardinal), Roan (a mixture of colors), and Rouen (a town in north central France). Eventually the name Rouen was settled on both in England and France.

With their amazing proclivity for producing and refining breeds of poultry, the British soon redesigned the Rouen. They altered its sleek "puddle duck" form into a thickset boat shape, doubled its size, and improved its colors. The Rouen was used primarily as a high-quality roasting duck when butchered at 5 to 6 months of age.

According to Paul Ives in his book *Domestic Geese and Ducks* (1947), Rouens were imported into the United States by D. W. Lincoln of Worcester, Massachusetts, in 1850. In North America, the Rouen became popular as a colorful general-purpose farm duck, and prior to the arrival of the Pekin was used for the production of market ducklings.

The Rouen was included in the first APA *Standard of Perfection* in 1874, and ever since has been regarded by many to be the ultimate exhibition duck for its beauty, size, and the challenge involved in breeding truly good show specimens. At major shows it often accrues the most entries among the Heavyweight Class and is a frequent winner in competition with other breeds.

Description

There are two distinct types of ducks raised in North America that are known as Rouens. The common or production-bred Rouen is a Mallard-colored bird with "normal" duck conformation, generally weighing 6 to

A pair of production-bred Rouens. The drake has just begun the eclipse molt, as evidenced by the sprinkling of dark feathers on his gray underside.

8 pounds. These are raised primarily as general-purpose ducks and for decoration.

The standard-bred Rouen is a huge block-shaped duck weighing 9 to 12 pounds with the most highly perfected Mallard coloration of any breed. These birds are bred principally for exhibition and decoration, but make great pets and large, tasty roasting ducks. In size and type, they are almost identical to the standard-bred Aylesbury, except for the shape of the head, bill, and back. In Rouens, ideally the head is rounder, the bill a bit shorter and more concave along the top line, and the back shows more arch from the shoulders to the tail.

Varieties

Gray is the original color. In the past few years, Black, Blue Fawn, and Pastel varieties have been developed and exhibited.

Selecting Breeders

In production-bred strains, the selection criteria should be based on the characteristics that best fit your needs.

A massive standard-bred Rouen old duck bred by Danny Padgett of Florida and owned by Gene Bunting of Oregon.

In the standard-bred Rouen, emphasis should be placed on vigor, strong legs, large size, good body length, deep keels that are straight and level, horizontal body carriage, and proper color and markings. Sometimes the largest and deepest-keeled drakes have poor fertility, so it can pay to retain for breeding some males that are not quite as extreme in size and keel development. The arch in the back of a Rouen normally becomes more pronounced with age, so one must be cautious in breeding from young birds with excessively arched backs.

Selecting and Preparing Show Birds

In good competition, the winners have excellent size, long bodies displaying good depth, straight keels that run from the breast to the abdomen, smoothly arched backs, fine color and markings, and sparkling condition. To get them in good show condition, Rouens should be kept in clean pens with fresh bathing water and sufficient room for exercise (I use pens at least 10 feet long). Rouens should be fed either free-choice or twice daily a quantity that they clean up in 10 to 20 minutes for 6 to 8 weeks prior to and throughout the show season, with a ration consisting of 3 parts gamebird flight conditioner, 1 part cat kibbles, and 1 part oat pellets or whole heavy oats, to help promote keel development and fine feather condition.

Comments

With deep-keeled Rouens, fertility is usually improved if swimming water at least 4 to 6 inches deep is available. Standard-bred ducks are so heavy that they tend to crush their eggs if allowed to incubate them, but some production-bred females are good setters. Egg color is usually white, tinted, green, or blue.

Saxony

In eastern Germany, about 25 miles from the Czechoslovakian border, Albert Franz of Chemnitz began working in 1930 on creating a new multipurpose duck. He used Rouen, German Pekin, and Blue Pomeranian in his breeding program. In 1934, his creation was introduced at the Saxony Show. Unfortunately, few Saxony remained at the end of World War II, so Franz renewed his breeding program. The Saxony was recognized as an official breed in Germany in 1957, where it continues to be esteemed for its enchanting beauty and the production of large eggs and full-breasted roasting ducks of superior quality.

The Saxony was introduced in the United States in 1984. Their numbers have been increasing as they gain recognition as a hardy breed that combines unique plumage color with excellent practical qualities. A qualifying meet for admittance into the APA *Standard of Perfection* is scheduled for the fall of 2000 in Oregon.

Description

In type and size, the Saxony resembles the Appleyard. It is a moderately large and sturdily built duck that does not have the extreme coarseness of features nor the loose feathering of the Pekin. Due to their tight plumage and thick muscling, they are often heavier than they appear.

Their bill is stout, of medium length, and with a nearly straight top line. The moderately large head is fairly oval in shape, with only a slight forehead crown. The head blends smoothly into a medium-length neck that is moderately thick and slightly arched forward. The compact body is moderately long, broad across the shoulders, with a prominent chest that is smoothly rounded. The underbody is smooth and the abdomen moderately full. The back has good width and is nearly straight along its top line, with gently rounded shoulders. The tail is carried slightly above the line of the back. Body carriage is approximately 25 degrees above horizontal when the bird is relaxed, but may go up to 40 degrees when excited.

Varieties

The name Saxony refers to both the breed and its single variety.

Selecting Breeders

Along with the Appleyard, the Saxony is the most active forager and best layer among the large ducks of Mallard descent. Therefore, it is important to use breeders that are robust, strong-legged, active, and prolific layers of large eggs. Many Saxony are undersized, so heavily muscled birds of good size are valuable for breeding. On the other hand, excessively large size is not desired since it hampers foraging ability and laying efficiency.

A young pair of Saxony bred on our farm. The duck at left has the desired creamy white throat, neck front, and facial stripes, and the drake has the desired white neck collar that completely encircles the neck.

Selecting and Preparing Show Birds

The Saxony is a beautiful exhibition bird, and its unique colors provide a pleasing contrast to other breeds in the showroom. Moderately large, solidly built, no keel on the breast, silky smooth plumage, rich colors, and clearly defined markings are the main characteristics to look for in show birds.

The soft colors of the Saxony will fade and discolor under prolonged exposure to strong sunlight, so the birds should have access to ample shade. A diet consisting of 3 parts gamebird flight conditioner, 1 part oat pellets or whole heavy oats, and 1 part cat kibbles will promote well-fleshed bodies and fine feather condition.

Comments

Compared with Pekins, Saxony do not grow quite as fast, but they have much more interesting plumage, produce meat with more flavor and less fat, are better foragers, and are more likely to incubate and hatch their eggs. Most females lay large, white eggs. Saxony are one of the best large all-purpose breeds of ducks and adapt well to a wide range of environments.

Eight

THE IMPORTANCE OF PRESERVING RARE BREEDS

For thousands of years, people throughout the world have raised animals for companionship, work, food, clothing, pest control, beauty, and sport. Sometimes by chance but often by design, animals were naturally selected over time that were well suited for specific tasks or environments. Soon after the turn of the twentieth century, the number of breeds of domesticated animals approached its zenith. Farm animals, for example, were suited to an amazing array of environments and uses.

As early as the 1920s, a new trend began to emerge. The mechanization and specialization of farms ushered in an era when fewer and fewer breeds were relied on to produce food. The livestock and poultry industries focused increasingly on specialized breeds and hybrids that could be packed tightly in minimal space. Improved yields and greater efficiency became the rallying cry of the "modern" agriculturists. With uncanny speed, this tunnel vision led to near decimation, or in some cases total loss, of various old breeds.

By the 1960s, whispers of warning could be heard. Some stubborn farm folks who had turned a deaf ear to the experts insisted that the old breeds still had their place in the modern world. In the halls of academia, a few forward-thinking geneticists pointed out that the losses of genetic stocks of plants and animals, both wild and domestic, could have profoundly negative effects on the quality of life on planet earth, even decreasing humans' chances for long-term survival.

The Dutch Hook Billed has been traced to the 1600s and nearly became extinct before a few breeders in Holland, Germany, and England saved them in the later part of the last century. Imported to the United States in spring 2000, this 10-week-old pair were part of the first clutch of Hook Billed ducklings hatched in North America. This fascinating breed, now being bred and studied at our Waterfowl Farm, comes in white-bibbed (shown here), non-bibbed, and pure white.

How You Can Help

As we head into the twenty-first century, preserving the widest possible gene pool in all types of plants and animals is essential. Around the world, geneticists, trained plant and animal breeders, and — just as important — thousands of lay people are dedicated guardians of endangered genetic stocks of all kinds. A variety of organizations now promote this work, including Seed Savers and National Germplasm Repositories (for plants) and the American Livestock Breeds Conservancy, the Society for the Preservation of Poultry Antiquities, and International Registry of Poultry Genetic Stocks (for animals and fowl). Contact information for some of these is in appendix H on page 302.

Keep Rare Breeds

One of the most important contributions you can make is to actually keep one or more of the rare breeds. By purchasing endangered breeds, you make three important contributions: You support the farms that breed them, you increase the populations of the breeds, and you augment the number of locations where the breeds are kept, thus reducing the possibility that they could be wiped out by plague, marauding predators, or natural disasters.

Multiply Rare Breeds

A step beyond simply keeping rare ducks is breeding them. By propagating them you are increasing their numbers and providing yourself with additional birds from which to select future breeding stock. Plus, you can distribute birds to other interested persons.

If sufficient young birds are raised each year, the best specimens can be retained for your own breeding flock or distributed to other people, and the less typical specimens can be used for food, pets, or decoration. If the superior specimens are saved for breeding each year, the overall quality of the gene pool can be maintained or, better yet, improved.

Learn about Your Chosen Breeds

Learning the history, conformation, color, and other characteristics of chosen breeds can increase your enjoyment, improve your ability to select breeding stock, and make you a better rare-breeds promoter. When promoting rare breeds, remember that people are usually attracted to a particular breed for one or more of the following reasons: appearance, practical qualities, and/or history. Experience has shown that the more one is engaged with the story of a breed, the more likely one is to continue raising it.

Exhibit Your Rare Breeds

One of the best ways to introduce people to heritage breeds and create interest is to exhibit your rare birds at county, state, provincial, or poultry club shows. When showing rare breeds or varieties, a wonderful way to educate people is to prepare a poster/card that gives pertinent breed information and place it on or near the display cage of the birds. An example of the type of information card we have used is shown at the top of page 91.

Typed on paper, this card can be thumbtacked to a slightly larger piece of ¼-inch plywood, outfitted with two opened paper clips, and hung on a cage. You should ask the show superintendent for permission — in our experience

ANCONA DUCKS

Origin: Great Britain

Weight: 5 to 7 pounds

Color Varieties: Black, Blue, Chocolate, Lavender, and Tricolored

Practical Qualities: Excellent layers (210 to 280 per year) of white, cream, green, blue, or spotted eggs; high-quality meat; excellent foragers that eat large quantities of slugs, snails, and insect pests; easy to see in field or on pond

Characteristics: The plumage pattern is broken with irregular white spots and patches throughout; seldom are two individuals marked exactly alike; the bill and feet are spotted

Status: Extremely rare

For more information contact: (your name, address, phone number)

superintendents have always been pleased with these information plaques because they add class and interest to a show.

An additional way to disseminate information at shows is to make a simple wooden holder for free flyers that describe your efforts (or others') to improve and preserve rare breeds. The holder with flyers can be placed in a prominent location, such as on top of the show cage.

Some rare breeds, due to their scarcity, have not been officially recognized by the American Poultry Association (APA), even if they are recognized in their country of origin. Some shows will accept competition entries only for the breeds and varieties officially recognized by the APA. However, most shows will accept entries of rare breeds even if they are not in the *American Standard*. Even those shows that do not accept nonstandard breeds for competitive classes normally will allow them to be entered for display.

Support Heritage Breed Organizations

There are a number of organizations that are dedicated to helping preserve rare breeds of livestock. They publish newsletters, sponsor seminars, and do a variety of promotional work. Supporting them with your membership can increase your knowledge and ability to work effectively for the good of heritage breeds. Appendix H features contact information for these and other organizations.

Help Standardize a Rare Breed

One step in helping increase the chances of survival of a breed is getting it recognized by the APA. Once a breed is officially standardized, there is more incentive for specialty breeders and exhibitors to raise and show it, which in turn increases the breed's exposure to the public. Currently the procedure for getting a breed standardized includes the following:

1. At least 5 people must raise the breed for a minimum of 5 years.

2. A standard description for size, shape, and color is prepared for the breed.

3. At the end of the 5 years, at least 5 exhibitors must show a minimum of 50 specimens at a qualifying meet.

4. The judge of the qualifying meet must recommend to the APA Board of Directors that the breed be admitted due to sufficient uniformity as well as conformity to the written standard.

5. The APA Board of Directors votes on admittance.

While the task of getting a breed officially recognized sounds a bit arduous, it is feasible when a group of breeders pitches in to help.

The primary purpose of conserving rare breeds is for the benefit of future generations. Maggie, our 3-year-old niece, delights in watching over these day-old Saxony ducklings.

NINE

HYBRID DUCKS

Sometimes, for special purposes, hybrid ducks are produced by mating together drakes and ducks of different breeds. Possible advantages of this scheme include: the ability to combine characteristics that are not available in any one breed; the positive effects of heterosis (often referred to as *hybrid vigor*); and the use of sex-linked matings that produce offspring that can be sexed by color at hatching.

Heterosis

Heterosis refers to the difference in growth and production traits between crossbred offspring and the average of their parents. An example is the mating of a drake from a breed that lays 300 eggs a year to a duck of a breed that normally lays 100 eggs a year. It would be logical to expect their offspring to produce 200 eggs per year. If, in fact, their female progeny produced 285 per year, the 85 "bonus" eggs could be credited to the positive effects of heterosis. A side note: Heterosis sometimes has a negative effect on crossbred offspring, a fact that is sometimes overlooked. For example, when two breeds are crossed, their offspring often will have slightly higher body fat than the average of their parents. If your goal is to produce meat with a minimum of fat, this would be an example of a negative effect of heterosis.

What Can Be Crossed

All breeds of domestic ducks can successfully cross if there is not too great a difference in the lengths of their bodies. Generally, the breeds in the Light-, Medium-, and Heavyweight Classes can mate naturally with each other.

However, the greater the difference in body size between the breeds being crossed, the lower fertility will usually be. It is difficult for the Bantam breeds to cross with larger ducks. Artificial insemination is sometimes employed to improve fertility or to make crosses between breeds that would be difficult or impossible for the birds to accomplish naturally.

Disadvantage of Hybrids

A major disadvantage of hybrids is that they often do not make suitable breeders. Some hybrids are sterile. When fertile hybrids are mated together, their progeny display great variability in their physical and production traits and a portion of heterosis is lost. Therefore, to produce hybrids, new pure-bred breeding stock is required each generation.

Two-Way Hybrids

The most common method for producing hybrid ducks is to use a two-way cross. Drakes from breed (A) are mated to ducks from breed (B) to produce hybrid offspring (AB). Or, the reciprocal cross can be made in which drakes from breed (B) are mated to ducks from breed (A) to produce hybrid off-spring (BA).

If you decide to produce hybrid ducks, it is important to understand that the offspring that will result can look and perform differently depending on which breed is used on the male and female sides of the cross. For example, when Khaki Campbell drakes from a high-producing strain are crossed with Black Cayuga ducks, their sons will be black, whereas the daughters will be chocolate and capable of laying approximately 300 eggs per year. In the reciprocal cross of Cayuga drakes x Khaki Campbell ducks, all of the offspring are black and the daughters are capable of laying about 250 eggs per year.

Three-Way Cross Hybrids

To combine the traits of three breeds or strains, drakes from breed (C) are mated with hybrid ducks (AB) to give progeny (CAB). Or, the reciprocal cross, drakes (AB) mated to ducks from breed (C) to give progeny (ABC), can be used.

These three-way white hybrids were produced by mating White Campbell drakes to Pekin ducks. The F₁ offspring were then mated to White Runner ducks. The resulting offspring are quick growing, well muscled, and significantly leaner than Pekins, and the daughters are durable layers of white- or blue-shelled eggs.

Four-Way Cross Hybrids

To combine the traits from four different breeds or strains, you first need to cross males (A) × females (B) to produce progeny (AB) and males (C) × females (D) to produce progeny (CD). In the second generation, the (AB) birds are mated to (CD) birds to produce (ABCD) offspring. At each stage, a reciprocal crossing can be made.

Sex-Linked Hybrids

There are situations when it is useful if ducklings can be sexed simply by looking at the color of their down, a phenomenon commonly referred to as sex-linked coloring. Sex-linked matings are made by crossing any brown-colored drake that carries two brown dilution (d/d) or two buff dilution (bu/bu) genes onto females that are genetically black or gray. In their offspring, the colored portions of the down will be black or near black in males and brown in females. The reciprocal cross of gray or black drakes × brown ducks is not sex-linked and will produce all gray or black progeny.

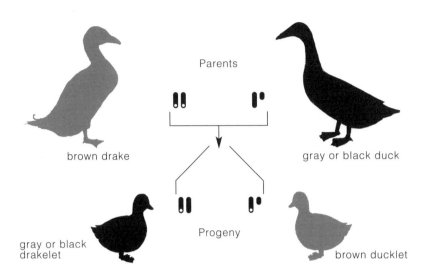

A brown drake X *a gray or black duck is a sex-linked mating that produces gray or black sons and brown daughters.*

Brown-colored drakes that can be used for sex-linked matings include Khaki Campbell, Penciled Runner, Chocolate Runner, Golden Harlequin, Chocolate Magpie, Chocolate Ancona, and Chocolate Muscovy. (Fawn & White Runner, Buff Runner, and Buff Orpington drakes can also be used, but they carry blue dilution, which makes it more difficult to distinguish the sexes at hatching time, and therefore they are not recommended.)

Females that can be used for mating to the brown drakes include Dark Campbell, Black Runner, Gray Runner, Silver Harlequin, Black Magpie, Black Ancona, Black Cayuga, Black Swedish, Silver Appleyard, Gray Rouen, and Black Muscovy. (Technically, any blue-colored female can be used as well, but this is also not recommended.) Most Pekins carry gray or black "underneath" their white plumage (the expression of their true color is prevented by the presence of two "white" genes) and therefore can be used on the female side of a sex-linked cross.

At a poultry research station in the Caribbean during the 1970s, my students and I studied many sex-linked crosses as we searched for combinations that might produce useful birds for subsistence farmers in developing countries. We found that any cross using Khaki Campbell drakes out of high-producing strains produced marvelous laying ducks, no matter what breed their mates were (excluding Muscovy).

Khaki Campbell x Black Runner produced 3.5- to 4-pound hybrids that laid 290 to 320 eggs per hen per year and had unsurpassed foraging ability. Khaki Campbell x Cayuga produced 5.5-pound birds that laid 270 to 300 eggs per hen per year and were relatively calm and good foragers. Khaki Campbell x Pekin (a large commercial strain) produced progeny that weighed 6 to 7 pounds and laid 290 to 320 eggs per hen per year; these were the calmest of the crosses tried.

Harlequins, Runners, Magpies, and Anconas are also excellent sires of productive layers when crossed onto other breeds of Mallard descent. Harlequin-sired hybrids lay about as well as Campbell hybrids and their eggs are slightly larger. Hybrid daughters sired by Magpies lay the largest eggs.

Briefly, the reason sex-linked matings work is due to the location of the recessive brown dilution gene (d) on the sex chromosomes. Male ducks are the *homogametic sex* because they carry a matched pair of sex chromosomes designated as ZZ. Female ducks are called the *heterogametic sex* since their two sex chromosomes, designated ZW, are of different lengths (the W chromosome is shorter). (Note: In mammals, the male is XY and the female is XX.) The brown dilution gene (d) is located on that portion of the Z chromosome that is missing in the W chromosome. Therefore, a female duck can only inherit the brown dilution gene (d) from her father.

Muscovy Hybrids

When Muscovies are crossed with ducks of Mallard descent, the offspring are sterile "mules." This cross produces a wonderful meat bird that combines the fast growth rate of Mallard derivatives with the heavy muscling and leaner carcass of the Muscovy. The cross can be beautiful and exotically colored.

Commercially, Muscovy drakes are crossed on Pekin, Rouen, or other large ducks. Eggs hatch in 4 weeks, and males and females are a similar size at butchering age. The hybrid females usually lay small eggs that do not hatch.

The reciprocal cross of Mallard-derivative drake x Muscovy female produces offspring that take 5 weeks to hatch and have considerable size difference between the sexes, as well as unproductive females.

For the best fertility, use Muscovy drakes that have not been with Muscovy females since the age of 8 to 12 weeks, with one Muscovy drake for every five to eight ducks. In general, fertility for this cross runs at 40 to 60 percent when eggs are candled on the 7th day of incubation. Some commercial producers use artificial insemination.

TEN

UNDERSTANDING DUCK COLORS

A valuable tool of the duck breeder is a working knowledge of colors. Along with aesthetic appeal, colors have important practical consequences, including camouflage from predators, appearance of dressed carcass, temperature control (white or light-colored ducks reflect more of the sun's rays), feather durability, resistance to soiling, and breed identification.

The importance of color in breed preservation should not be underestimated. Some people argue that only practical qualities are important and color is superfluous. Throughout history there have been poultry breeders who ignored color and focused their breeding programs solely on "practical" qualities. Over time their lines of birds were lost due to the lack of easy-to-identify plumage "markers." It should also be remembered that the genes responsible for plumage appearance can also exert influence on "practical" qualities, such as disease resistance, growth rate, and fertility.

Basic Genetics

Having a rudimentary understanding of genetics is essential to understanding duck colors and knowing how to improve them through breeding. The following is a brief review of the genetic principles required to understand basic color inheritance in ducks. (For detailed information, study Mendelian genetics in a good textbook. Or better yet, audit or take a beginning genetics course at a local college or university.) Getting a grasp on the inheritance of colors takes time, study, and review — unfortunately many people give up before giving themselves a chance to absorb the information.

Chromosomes

At a basic level, genetics is the study of why an egg laid by a duck produces a baby duck rather than, say, a duckbill platypus. The reason why "like begets like" is due to the genetic blueprint carried on the threadlike chromosome found in the nucleus of each cell.

Ducks have 80 chromosomes arranged into 40 pairs. A duckling inherits one chromosome of each pair from its mother, the other from its father. There are 39 pairs of *autosomal chromosomes* in which the chromosomes of a given pair are matched for size. The 40th pair are the *sex chromosomes*. As mentioned in chapter 9, the male is the *homogametic sex* because it carries a matched pair of sex chromosomes designated as ZZ. The female bird is the *heterogametic sex* since with her two sex chromosomes, labeled ZW, the W is shorter.

Genes

Located along each chromosome are genes, which can be thought of as the codes or switches that control the characteristics of a duck. Some characteristics are controlled by a single pair of genes (one gene located on each of a matched autosomal pair of chromosomes), whereas other characteristics are controlled by multiple pairs of genes. The specific location where a gene is found on a chromosome is called the *locus* (the plural is *loci*). Genes are given shorthand symbols such as (Bl) for blue dilution and (c) for white.

Wild Type and Mutations

For any given trait, the original gene(s) found in the wild ancestor of domestic ducks is/are known as the *wild type*. Occasionally a gene (or chromosome) changes, which is known as a *mutation*. In this chapter, when gene symbols are used, the wild type will be designated by a $^+$. For example, the blue-dilution mutation is shown as Bl, whereas the nondilution (wild type) allele is bl^+.

Genetic mutations can affect size, color, plumage pattern, shape, disease resistance, skeletal formation, and more. Mutations can be harmful, neutral, or beneficial. Some mutations are harmful or even lethal in a double dose, whereas in a single dose they are neutral or beneficial. All domestic breeds of ducks, except those colored exactly like wild Mallards or wild Muscovies, carry at least one mutation gene for color. Mutated color genes can be added together in various combinations to make different colors and patterns.

Alleles

Alternative forms of a gene are called *alleles*. If a gene has one recognized mutation form, then it is said to have two alleles (the wild type plus the mutation form). If a gene has two recognized mutation forms, then it is said to have three alleles (the wild type plus the two mutation forms), and so on.

In literature, you can always tell which genes are thought to be alleles because they will have the same symbol but different superscripts. An example is the mallard pattern series of restricted (M^R), dusky (M^D), and mallard (m^+). As you can see, the symbol for these three alleles is the letter M, with different superscripts indicating which one of the series is being discussed. Capitalized letters are used for dominant alleles, while lowercase is used for recessives.

Homozygote and Heterozygote

If a duck carries a matched pair of genes at a particular locus, it is said to be a *homozygote* for that site, i.e., Bl/Bl. On the other hand, if it carries different alleles at a site, it is called a *heterozygote*, i.e., Bl/bl$^+$. Breeders and hobbyists commonly use the terms "split" or "carrier" for birds that are heterozygous for color. For example, when attempting to improve the type of Snowy Calls by crossing onto Gray Calls with outstanding conformation, the first-generation (F_1) offspring would be phenotypically Grays, but would be described genotypically as "Snowy carriers" or "split to Snowy." (See discussion of phenotype and genotype below.)

Ducks are said to "breed true" when all of the offspring resemble the parents from generation to generation. In order to breed true, both parents must be homozygous and have identical genotypes for the traits under consideration.

Phenotype and Genotype

Phenotype refers to the outward appearance of an individual, whereas *genotype* is the actual genetic makeup of the individual. The phenotype of a duck's plumage is not always an accurate barometer of its genotype. For example, when you cross a Snowy Call with a Gray Call, the first-generation offspring normally look like Grays and often give no clues that they "carry" Snowy. So the phenotype of these crossbreds is Gray but their genotype includes one gene for Gray and one gene for Snowy (Li^+/li^h). Mate these crosses together and one out of four of their offspring will be Snowy.

When genotypes other than wild type are given in this chapter, only the *known* mutation genes will be identified. For example, the genotype for ducks with solid black plumage is given as E/E.

Dominant and Recessive Genes

Genes normally occur in pairs (one located on each matched chromosome). If a duck carries one wild-type gene and one mutation gene at a particular locus (site), and the color of the bird is not changed, then the wild type is said to be *dominant* and the mutation gene *recessive*. For example, when a Gray Mallard (C^+/C^+) is crossed with a White Mallard (c/c) the offspring will be colored and will show no indication that they carry one white gene (C^+/c). Gray is said to be *completely dominant* to white. Despite being normal mallard-colored, these heterozygous (C^+/c) ducks are white "carriers," and when mated together typically will produce offspring in a ratio of 3 Grays: 1 White.

The relationship of two genes at the same locus is not always so clear-cut. If a duck carries one wild-type gene and one mutation gene at a locus and both alleles affect the appearance of the bird, then the mutation gene is said to be *incompletely dominant* to wild type. Blue dilution (Bl) is an example of an incompletely dominant mutation gene. If a blue-dilution gene is added to a Gray Mallard, the resulting color is Blue Fawn (Bl/bl^+), whereas the addition of two blue-dilution genes results in Pastel (Bl/Bl).

Sex-Linked Genes

If a gene is located on the "bottom end" of the Z chromosome (the segment "missing" on the W chromosome), then a female duck can only inherit it from her father and it is called a *sex-linked* gene. Both sex-linked and recessive genes can cause characteristics to "disappear," only to reappear in a later generation. For this reason, experimental breeding programs should be continued for a minimum of two generations to know the true genotype.

When genotypes are given in this book, the second symbol for any sex-linked gene pairs will be italicized to indicate that females carry a single gene at that locus. For example, the genotype for khaki colored birds is symbolized as M^D/M^D, d/*d*.

Genetic Ratios

In discussing color genetics, various ratios are mentioned. For example, when a blue drake is mated to a blue duck, their offspring will hatch in a ratio of 1 Black: 2 Blue: 1 Silver or Silver Splash. These ratios are usually accurate when hatching hundreds or thousands of offspring. However, when hatching small numbers of ducklings, do not be surprised if these ratios are off by a fair amount.

Mallard Derivative Colors

The American Poultry Association (APA) recognizes 15 color varieties in ducks of Mallard descent. This section will concentrate on these varieties, plus Ancona, Australian Spotted, Harlequin, and Saxony. (For detailed descriptions of the Standard varieties, see the latest edition of *The American Standard of Perfection*, published by the APA.)

The plumage colors and patterns found in the 19 varieties discussed here are the result of *wild-type* genes, plus 11 *main mutations* as outlined in the table below. The color of non-wild-type varieties is caused by anywhere from a single mutation to a combination of as many as five mutations.

In addition to the genes listed in the table, keep in mind that there are "modifying genes" that fine-tune plumage colors and patterns. These secondary genes are difficult to study and in general are poorly understood. The actions of these modifying genes are of special interest to breeders interested in improving their purebred stock.

To help understand how duck colors are "built," the varieties can be divided into six categories based on their genetic foundation:

1. Mallard Pattern (Gray, Pastel, Saxony, Snowy, and Spotted)
2. Restricted Pattern (Silver Appleyard)
3. Dusky Pattern (Khaki, Harlequin, and Buff)
4. Spread Black (Black, Blue, and Chocolate)
5. Self White
6. Pattern White (Bibbed, Fawn & White, Penciled, Magpie, and Ancona)

Plumage Color and Pattern Genes
in Mallard-Derivative Ducks

NAME	SYMBOL	RELATIONSHIP TO "WILD"	MAIN VISUAL EFFECTS ON ADULTS
Restricted Pattern	M^R	Dominant	Whitens surface of wing fronts; slightly lightens overall plumage
Dusky Pattern	M^D	Incompletely dominant	Eliminates facial stripes, obscures wing speculum; in drake eliminates neck ring and claret breast
Mallard Pattern	m^+	Wild type	Wild Mallard pattern
Dark Phase	Li^+	Wild type	Wild Mallard coloration
Light Phase	li	Recessive	Moderately lightens plumage; in drakes extends claret onto shoulders and sides; sometimes enlarges neck ring
Harlequin Phase	li^h	Recessive	Further lightens plumage; enlarges wing speculaum; in drakes extends claret onto shoulders and sides, darkens main tail feathers, ring often encircles neck
Colored	C^+	Wild type	Permits normal plumage color expression
White	c	Recessive	Prevents normal plumage color expression
Extended Black	E	Incompletely dominant	Extends black throughout each colored feather
Non-black	e^+	Wild type	Permits normal plumage color expression
Blue Dilution	Bl	Incompletely dominant	Dilutes black pigment to blue gray in single dose; silver or silver-splashed white in double dose
Non-blue Dilution	bl^+	Wild type	Permits normal plumage color expression
Non-brown Dilution	D^+	Wild type	Permits normal plumage color expression
Brown Dilution	d^d	Sex-linked recessive	Dilutes black pigment to dark brown
Non-buff Dilution	Bu^+	Wild type	Permits normal plumage color expression
Buff Dilution	bu	Sex-linked recessive	Dilutes black pigment to medium brown
Runner Pattern	R	Incompletely dominant	Prevents pigmentation of neck and parts of head, wings, and underbody
Non-runner Pattern	r^+	Wild type	Permits normal pigmentation throughout plumage
Dominant Bib	S	Incompletely dominant	Prevents pigmentation on front of chest
Non-bib	s^+	Wild type	Permits normal pigmentation throughout plumage

The Four Plumages of Ducks

Ducks of Mallard descent have four distinct plumages during their life. Ducklings are covered with fur-like *down.* At eight weeks of age, most of the down has been replaced with *juvenile* feathering, and both genders resemble adult females.

By 16 to 20 weeks of age, young ducks sport their dimorphic *nuptial* or breeding plumage, which is the plumage described in breed standards. At the end of each breeding season, both sexes molt into a juvenile-like *eclipse* plumage that is retained for approximately four months.

At 17 weeks of age, this Mallard drake has begun to replace his juvenile plumage with the bright colors of adulthood. To provide camouflage as long as possible, young males acquire their iridescent heads and white neck collars last.

A standard-bred Mallard drake in full nuptial plumage in early spring.

The same drake as shown above, but 4 months later in full eclipse plumage. By fall, he will again sport flashy nuptial plumage.

Gray

The wild mallard pattern is officially known as Gray in North America. All other varieties are mutations of the Gray. For show purposes, the gray color variety is the most perfected in standard-bred Rouens.

Breeds
Call, Mallard, Runner, Rouen

Genotype
m^+/m^+, Li^+/Li^+, C^+/C^+, e^+/e^+, bl^+/bl^+, D^+/D^+, Bu^+/Bu^+, r^+/r^+, s^+/s^+

Day-Old Description
Ducklings have the classic black and yellow camouflage pattern. Greenish black runs from the base of the bill, over the top of the head, down the back of the neck, and spreads out across the shoulders, back, and tail. The face is yellow with one or two dark facial stripes and the back and wings are marked with yellow "sun spots." The front of the neck, chest, and underbody are yellow. The bill is black, often with yellow shading on the margins. The feet and legs are gray yellow with dark webs. When these ducklings hold still in grass, they become nearly invisible as their colors blend in with sunspots and shadows.

Adult Description
The drake in nuptial plumage has an iridescent green head, white neck collar that does not meet in the back, claret chest, steel-gray underbody, black rump and under-tail coverts, white to ashy-gray tail feathers, and reddish-orange feet and legs. Depending on the time of year, the bill varies from yellow to olive green with a black bean tip.

The subdued tones and intricate pattern of the duck give her superb camouflage. Her base color is golden brown to rich mahogany, with each feather distinctly penciled with dark brown. The legs and feet are brownish orange and the bill is orange to brownish orange, with a dark saddle across the middle and a black bean at the tip. Both sexes have bright-blue wing speculums and brown eyes.

Juvenile drakes resemble females, but can often be identified by their green bills and darker color with less penciling on the top of the head and lower back. The bright colors of the nuptial plumage gradually begin appearing at 10 to 12 weeks and are completed by 16 to 20 weeks of age. Juvenile females have duller, less distinctly penciled plumage than adults. For several months each year (usually in late spring and summer), drakes exchange their bright hues for subdued, female-like eclipse plumage as they replace their flight feathers.

Common Faults

For exhibition, the following are undesirable in drakes: clear yellow bill, foreign color on the head, poorly defined white neck ring, light lacing on claret feathers of the chest (this "chain armor" typically is most noticeable on newly grown feathers and then darkens), muddy steel gray on body, white or gray mixed in with the black of the under-tail coverts, and black on the ridge of bill (most Calls have this fault).

In the duck — as much as possible — avoid: indistinct facial stripes, poor penciling in body plumage and/or wing bows, faded color on the underbody and under-tail coverts (usually, even the best exhibition Call females are pale in under-tail covert color), solid white feathers on front of neck, and solid lead-colored bills (due to hormonal changes, it is normal for the bill to darken prior to and during the laying season). In both sexes, watch for poorly colored wing speculum with indistinct black and white border on top and bottom.

Breeding Hints

In exhibition Rouen drakes, it is preferred that there not be a vertical white border of feathers between the steel gray of the flank and the black of the under-tail coverts and rump. However, males that have this "cotton" as well as "chain armor" lacing on their chests normally produce daughters with the most distinct penciling. Also, drakes with the most distinct penciling in their juvenile plumage usually produce daughters with strong penciling.

Pastel

This soft color is called Apricot in Great Britain. Genetically, it is gray mallard pattern except for two blue-dilution genes.

Breeds

Call, Mallard, Runner, Crested, Rouen

Genotype

Wild type except for Bl/Bl

Day-Old Description

Pastel ducklings have the same pattern as Grays, but all black sections of the down are diluted to silver. The bill is grayish brown, and the legs and feet grayish yellow to brownish yellow.

Adult Description

Pastel drakes have the same pattern as Grays, but the head, neck, rump, and under-tail coverts are diluted to soft powder blue and the body is pale silver gray with a slight buff or rose hue. Depending on the time of year, the bill varies from yellow to green.

Overall, the duck is a blend of silver, buff, and salmon, often with a rose hue. The top of the head, back of the neck, and eye stripes are silver or brownish gray and the bill is orange with a dark-brown saddle. In both sexes, the wing speculum is bluish gray, bordered on top and bottom with white, and the legs and feet are orange to salmon.

Common Faults

Pastels can have the same pattern faults as the Grays. Drakes often have foreign color in their heads and some ducks have nearly solid silver bodies without the desired buff, salmon, and rose overtones.

Day-old exhibition Rouens in three colors, from left to right: Pastel, Blue Fawn, and Gray.

Breeding Hints

If you need to improve type in Pastels, mate them to Grays with excellent conformation and rich brown plumage. (Dark-colored Grays produce Pastel females with nearly solid silver bodies.) Pastel x Gray produces all Blue Fawn offspring. For the second generation, mate together the best typed Blue Fawns with rich brown tones in their juvenile plumage, and they will produce offspring in the ratio of 1 Gray: 2 Blue Fawn: 1 Pastel. When mated together, these Pastels will breed true.

Saxony

This soft color is named after the German Saxony breed. Genetically, it is gray mallard pattern except for two blue-dilution and two light genes. A common mistake people make is to assume that Saxony is a variant of the Buff color. Genetically, they are distinct colors that have no relationship to one another other than they both carry blue dilution.

Breeds

Call, Runner, German Saxony

Genotype

Wild type except for Bl/Bl and li/li

Day-Old Description

Saxony ducklings are similar to Pastels, but overall are lighter and a bit yellower in color.

Adult Description

At first glance, the Saxony and Pastel look similar. Upon closer examination, there are a number of distinctions.

In Saxony drakes, the claret breast color is larger and bleeds back onto the shoulders and sides of the body, and is frosted or laced with cream, especially on the lower portion. The oatmeal-colored body is paler than in the Pastel. Ideally, the Saxony drake's neck ring fully encircles the neck, although most drakes have a slight gap at the back. The head, neck, rump, and under-tail coverts are diluted to a soft powder blue. The bill is yellow, with varying amounts of green shading depending on the time of the year.

The plumage of the Saxony duck is a warm fawn to cinnamon buff with distinct creamy white facial stripes, throat, and neck front. The bill varies from yellow to brown, depending on the season of the year. Both sexes have silver-gray wing speculums bordered on top and bottom with white, and orange feet and legs.

Common Faults

Foreign color mixed in with the powder blue of the head and neck is especially a problem in older drakes. Brown-colored heads, indistinct white neck rings, and undersized claret bib on the chest in males are highly objectionable; in females, lack of creamy-white throat, neck front, and facial stripes are unacceptable in shows.

Breeding Hints

When choosing breeders, give priority to males with clean head color, distinct white neck rings, large claret bibs, and bills that are as yellow as possible. In females, look for warm fawn-buff color overall, offset by distinct creamy white on the throat, neck front, and face. Using some extra dark drakes and rich cinnamon-fawn females in the breeding pen can help restore faded overall coloring.

Comments

There are two forms of pseudo-Saxony coloration that can wreak havoc if they are mistakenly used in the breeding pen of true Saxony. The genotype of Type I pseudo-Saxony is M^D/M^D (occasionally M^D/m^+ or m^+/m^+), li/li, Bl/Bl, d/d, bu/bu, and is produced as a color variant by some strains of Buff Orpingtons. Phenotypically, these drakes often have poorly defined claret bibs, muddy white neck rings, and brown- or fawn-colored heads. Ducks either have solid fawn heads and necks or poorly defined facial stripes.

Type II pseudo-Saxony have a genotype of M^D/M^D, li/li, Bl/Bl, which is found in Runners and crossbred "pond" ducks from time to time. Drakes can closely mimic true Saxony, but often have poorly defined claret bibs and muddy white neck rings, while ducks have solid-colored fawn heads and necks.

These pseudo-Saxony colorations are due to the fact that when a pair of light genes are combined with dusky (M^D/M^D, li/li), the claret breast and white neck ring of the drake are at least partially restored. These imitation Saxony are sometimes promoted under the label of "American Saxony."

Snowy

This high-contrast color pattern is called "Silver" in Great Britain. Genetically, it is gray mallard pattern except for two harlequin genes. (Some birds labeled as Snowy are actually Harlequins since they carry the dusky pattern.)

Breeds
Call, Mallard

Genotype
Wild type except for li^h/li^h

Day-Old Description
These yellow ducklings look like they've been air-brushed with coal dust. The gray blush is usually heaviest on the head, lower back, and tail. For the first couple of days, ducklings can be sexed with 80- to 90-percent accuracy by bill color — most drakelets have gray or dark-green bills, while the majority of ducklets have yellow or tan bills with a dark tip. Legs and feet are gray or green with yellow shading.

Adult Description
Snowy plumage displays considerable variation in its expression among individual birds. Both sexes resemble their Mallard counterparts, except most of their body feathers are heavily edged with white. In drakes, the white neck collar fully encircles the neck, the heavily frosted claret coloration extends onto the shoulders and sides of the body, and the main tail feathers are smoky black with a narrow white border. The bill is greenish yellow to green, depending on the time of year. The feet and legs are orange shaded with brown.

In ducks, the head and neck are fawn, stippled with brown (especially on the crown of the head). The body is heavily frosted with white, each feather center-marked with brown. Bills are brownish orange with a dark saddle, and the feet and legs are orange with brown shading. Ideally, in both sexes, the brilliant blue wing speculum is enlarged, spilling onto the tertial feathers. Both sexes go through an eclipse molt, during which their plumage typically darkens.

Common Faults

In drakes, insufficient white frosting on chests, shoulders, and sides of bodies, neck collars that are narrow or do not fully encircle the neck, and light-colored tails are undesirable for exhibition. In ducks, common faults are nearly white heads and necks lacking the desired fawn base color and excessively white or dark bodies lacking sufficient frosting. In both sexes, dull or small wing speculums (it is fairly common for older females to lack brilliance in the speculum) are exhibition faults.

Breeding Hints

To improve type in the Snowy, outcross onto Grays with excellent conformation. To increase fawn coloring in the female Snowy, outcross onto Gray females with the richest brown or mahogany tones (avoid dark-colored Grays). Snowy × Gray crosses normally produce all Gray F_1 offspring. Mate the best typed F_1 males and females and they will produce offspring in the ratio of 3 Grays: 1 Snowy. These F_2 Snowys will breed true.

Spotted

This color is named after the Australian Spotted. The offspring of Spotted segregate out into three distinct plumage patterns: spotted, snowy, and light mallard (sometimes called Aleutian). Genetically, the spotted individuals appear to have the mallard pattern plus single genes for light phase and harlequin phase. Unnamed genes from the Northern Pintail and a wild Australian species may also be present (see chapter 4 for the history of the Australian Spotted). Australian Spotted are being bred in Greenhead, Bluehead, and Silverhead varieties. The Blueheads have a single blue-dilution gene and the Silverheads have a pair of blue-dilution genes.

A historical note: Some old-time breeders have told me that Snowy, Spotted, and Aleutian Calls were developed from the Australian Spotted.

Breeds

Call, Australian Spotted

Genotype

Greenhead: wild type except for li/li^h, and possibly others
Bluehead: wild type except for li/li^h, Bl/bl$^+$, and possibly others
Silverhead: wild type except for li/li^h, Bl/Bl, and possibly others

Day-Old Description

Greenhead ducklings resemble wild Mallards, with a bit more yellow in their down. In Blueheads the black portions of the down are blue gray, whereas in the Silverheads the black sections are silver.

Adult Description

Greenhead drakes resemble Mallards except the head is often marked with more or less brown and gray, the white collar fully encircles the neck, the claret of the breast extends back onto the shoulders and sides of body, and the underbody is a pale creamy silver that extends up onto the front of the chest. Depending on the time of year the bill is yellowish green to green. Drakes in eclipse plumage display conspicuous black or brown spots on most of their body feathers.

Greenhead ducks are light to medium fawn (the underbody, lower back, front, face, neck, and a faint neck ring are paler) with most feathers having an elongated dark brown center-mark. On the head, dark-brown stippling is most prominent across the crown and in a stripe through the eyes. The bill is orange and black. Both sexes have iridescent blue wing speculums and orange legs and feet with varying amounts of gray shading.

In Blueheads, all black and dark-brown portions of the plumage are diluted to blue gray; whereas in Silverheads, all black and dark-brown portions of the plumage are diluted to silver.

Common Faults

Due to the genetic makeup of this color, it is normal for there to be a range in the appearance of Spotted colored ducks. Common faults include indistinct white neck ring and lack of claret on the shoulders and sides of the body of drakes, and excessively dark underbodies and indistinct spotting in the body plumage of ducks.

Breeding Hints

For breeding, choose drakes with extensive claret on their shoulders and body sides when in nuptial plumage and bold spots in their juvenile and eclipse plumages, and ducks with distinctly spotted feathering. The subvarieties can be mated in any combination with the following results:

1. Greenhead × Greenhead produces all Greenhead offspring
2. Greenhead × Bluehead produces half Greenheads and half Blueheads

3. Greenhead × Silverhead produces all Blueheads
4. Bluehead × Bluehead produces all three colors in a ratio of 1 Green: 2 Blue: 1 Silver
5. Bluehead × Silverhead produces half Blueheads and half Silverheads
6. Silverhead × Silverhead produces all Silverheads

Silver Appleyard

This high-contrast color pattern is genetically wild type except for restricted pattern and two light genes. This is the only recognized duck color that carries the restricted mutation.

Breeds
Silver Appleyard

Genotype
Wild type except for M^R/M^R and li/li

Day-Old Description
Ducklings have smoky-yellow down with a black "mohawk" crown and black lower back and tail. Their bills are yellow with an irregular black stripe down the middle and feet are yellow with more or less gray shading.

Adult Description
Drakes resemble Gray Mallards with the following differences: silver-white throat and face flecked with fawn, white neck collar that fully encircles the neck, claret chest color lightly frosted with white that extends onto shoulders and along sides of body, and pale silver and cream underbody. In contrast to the snowy pattern, Appleyards have main tail feathers that are white, shaded with pale gray. Bills are greenish yellow, and legs and feet are orange.

Females are light-mallard colored, with white (preferred) or pale fawn extending from the throat, down the front and sides of the neck, onto the chest and underbody, and back to the abdomen. Running from the top of the bill across the crown of the head, down the back of the neck, and spreading across the shoulders and onto the sides of the body is an unbroken band of fawn feathers, center-marked with dark brown. The face is white and light fawn with a faint eye stripe. The surface of the wing fronts is

Day-old Silver Appleyards with typical dark tails and "mohawk" head markings against a yellow background. The dark head and tail markings are not this distinct in all Appleyard ducklings.

white. Bills are orange to brownish orange with a dark blotch on the ridge (they typically darken during the laying season) and the legs and feet are orange to brownish orange.

Common Faults

In drakes, smoky-black main tail feathers and solid-black throats are serious faults, indicating that they are not genotypic Appleyards. In ducks, excessive color on the throat, neck front, chest, and underbody; lack of white surface on wing fronts; fawn band down back of neck broken with white; and greenish-black bills are faults to avoid for exhibition birds. Many females with the overall best markings have dull or small iridescent wing speculums and should not be severely penalized for breeding or showing.

Breeding Hints

Drakes with the most silver white in their head plumage produce daughters with the best white-fronted color pattern. When possible, avoid breeding from drakes with solid-green heads and smoky-black tails, and ducks with solid-green or -black bills.

Khaki

This warm color is called Brown in some countries. Genetically, it is dusky pattern with brown dilution. A genetic quirk of the dusky pattern causes some offspring to have more or less white on the front of the neck and under the bill. Ducks of this variety do not have true khaki color until their plumage is faded from exposure to the elements.

Breeds
Call, Runner, Campbell

Genotype
Wild type except for M^D/M^D and d/d

Day-Old Description
Ducklings have olive-brown down with greenish- or bluish-brown bills and brown feet and legs.

Adult Description
In drakes, the head and neck, the lower back, wing speculum, and under-tail coverts are brownish black with a hint of green iridescence and the body is an even shade of medium- to tannish brown. Depending on the time of year the bill is greenish-yellow to green (some production-breds have bluish bills) and the legs and feet are orange. In ducks, the body is brown (with varying amounts of penciling), the head and neck are a shade darker than the body, the bill is blackish shaded with green, brown, or blue, and the feet and legs are brown to orange brown. (For exhibition, drakes with bronze heads and green bills are preferred, as are ducks with a minimum of penciling in their body plumage and wing speculums that match the body color.)

Common Faults
Ducklings or adults that have facial stripes are crossbreds and should not be used in a purebred breeding program or for exhibition. Also, avoid drakes with distinct white neck collars and claret on the chest.

Breeding Hints
To produce the best percentage of even-colored birds for exhibition, identify potential breeding drakes at 7 to 12 weeks of age that show the least penciling

in their juvenile plumage. Because white spreads, when possible avoid breeding from birds of either sex that have white on the neck or on the chin at the junction of the lower mandible and throat.

Harlequin

This high-contrast color pattern is named after the Welsh Harlequin breed. There are two color phases, the original Gold and the more common Silver. Genetically, the Gold phase is Khaki plus two harlequin genes; Silver-phase birds lack the brown dilution of the Khaki.

Breeds
Call, Welsh Harlequin

Genotype
Gold phase: M^D/M^D, li^h/li^h, d/d
Silver phase: M^D/M^D, li^h/li^h

Day-Old Description
Gold-phase ducklings are yellow, with more or less brown blush on the head and tail. Silver-phase ducklings are yellow with more or less gray blush on their head and tail (like Snowy ducklings). For the first couple of days after hatching, both phases can be sexed with about 90-percent accuracy by bill color: most drakelets have gray or dark-green bills, while ducklets usually have yellow or tan bills with a dark tip. (Some Harlequins carry modifying genes that lighten the bills of drakelets and darken the bills of ducklets.) Legs and feet are gray or green with yellow shading.

Adult Description
Silver-phase drakes closely resemble Snowys. Silver-phase ducks are light to medium fawn, frosted with white, and each body feather center-marked with dark brown. The head and neck are frosted white and shaded with light to medium fawn, similar to the body, but with darker brown stippling especially noticeable on the crown of the head. The bill is blackish green or blue and the legs and feet are orange brown. In both sexes, the wing speculum is bright blue and sometimes extends out onto the tertial feathers.

Gold-phase Harlequins are similar to the Silvers, but with slightly muted plumage colors. In Gold-phase birds, the head, neck, lower back, and under-tail coverts of the drake have a slight bronze hue; in ducks, the dark center-marks of the plumage are diluted to medium brown. In both sexes, the wing speculums are greenish bronze rather than the bright blue of the Silvers.

In both color phases, it is normal for males, especially those more than a year old, to have some silver and fawn flecking in their head color, especially around the eyes and in the feathering covering the ears.

Common Faults
In drakes, obscure white neck collars, indistinct white frosting or lacing on the feathers of the chest, shoulders, upper back, and sides of the body, and light-colored main tail feathers are faults to be avoided. In ducks, light-colored bills, head and neck color darker than the rest of body (this is a major difference between the Snowy and Harlequin — in Snowys, a darker head and neck color is highly desirable), body feathers lacking dark center-marks, head, neck, and body totally lacking fawn blush, and poorly defined wing speculum are common faults.

Breeding Hints
Diagnostic characteristics of Harlequins include black main tail feathers edged with white in drakes and no distinct facial stripe in ducks. Do not breed from drakes with light-colored tail feathers or ducks with obvious facial stripes.

Buff

This soft color is named after the Buff Orpington. Genetically, in North America it is dusky pattern with the addition of brown dilution, buff dilution, and two blue-dilution genes. In Great Britain, where they consider a darker cinnamon color to be ideal, their show birds have a single dose of blue dilution. Like khaki-colored ducks, some buff ducklings hatch with more or less white on the front of the neck and under the throat.

Breeds
Call, Runner, Orpington

Genotype
M^D/M^D, Bl/Bl, d/d, bu/bu

Day-Old Description
Ducklings have yellow-buff down, tannish-yellow bills, and orange feet and legs.

Adult Description
The ideal color that breeders strive for in both sexes is a rich fawn buff that is uniform throughout all surface areas of the plumage. However, even the best-colored drakes usually have head, neck, lower back, under-tail coverts, and speculums that are at least a shade darker than the rest of the plumage. Pale-colored wings are common.

Common Faults
Because buff-colored ducks carry blue dilution, they are predisposed to having blue-gray shading in their plumage. For exhibition, avoid birds that have strong blue overtones, distinct lacing, extremely light- or dark-colored plumage, and drakes with claret-colored breasts.

Breeding Hints
Some strains of buff-colored ducks are prone to produce females with facial stripes and drakes with claret on the chest. This problem can be eliminated by breeding from solid-headed females and drakes that have no facial stripes in their juvenile plumage.

Black

In solid-black breeds, the Cayugas and East Indies have the most perfected black plumage. Genetically, it is the result of two extended black genes. Due to genetic quirks of this color, many individuals have at least a bit of white in their plumage, and some day-old ducklings have sparse down on their backs.

Breeds
East Indie, Call, Runner, Magpie, Ancona, Crested, Cayuga, Swedish

Genotype
E/E

Day-Old Description
Ducklings have black down (often with a yellow blush on the chest and around the bill), bills, legs, and feet.

Adult Description
Ideally, both sexes have jet-black plumage with iridescent emerald sheen covering as much of the surface as possible, black bills (even the best-colored drakes usually have a bit of green or blue at the tip), and black to dusky black legs and feet (normally shaded with more or less orange in mature drakes).

The green iridescence of black plumage is produced by tiny prisms on the feathers. Because the sheen is produced by refracted light and not pigment, the quality of lighting under which black ducks are viewed greatly influences their color. (The best green color is obtained when black ducks are seen in diffused light consisting predominantly of medium to short wavelengths. Long wavelength light causes black ducks to have purple or bronze hues.) Also, black plumage that is worn loses iridescence and looks dark brown. Brilliant black plumage is restored after the bird goes through a complete molt.

Common Faults
For exhibition, avoid birds with brown lacing or penciling (check under the wing and on the throat), prominent white (some judges check under the throat), and strong purple sheen (even the best birds often have a bit of purple barring). Remember, in breeds with unusual conformation, such as the Runner, breed type is more important than color perfection.

Breeding Hints
To avoid unnecessary frustration, people who raise solid-black ducks need to understand that there are two types of white associated with them.

Juvenile white is usually present in the first plumage on the front of the neck or under the bill. Because juvenile white spreads quickly from generation to generation, it is not advisable to breed from individuals with prominent juvenile white. No matter how carefully breeders are selected, juvenile white cannot be totally eliminated from a strain of black ducks. (In fact, its presence is essential for hardiness and viability.) In carefully bred strains, 15 to 30 percent of offspring will display juvenile white.

Aging white is the equivalent of graying hair and is associated with green sheen. It normally shows up between 4 months and several years of age and is most prevalent in females. Aging white normally starts out as a few feathers with white edging, gradually increasing until the bird is mottled with white. Some ducks carrying extended black genes develop mostly white plumage with age, even if they were solid-colored as young birds. Typically, birds that display aging white produce offspring with the greenest plumage.

To produce offspring with the most iridescence, choose brilliant drakes that have green on the primary flight feathers and their underwing covert feathers, and mate them to ducks possessing good green sheen and that display at least a bit of aging white by the time they are 4 to 12 months of age.

To produce solid-black females that can be shown as old ducks, mate drakes and ducks that have no or a minimum of aging white at 2 years of age.

Blue

In Runners, this color is known as Cumberland Blue (commemorating the Runners first raised in Cumberland County, England). Genetically, it is extended black with a single blue-dilution gene. (Extended black and two blue-dilution genes produces plumage that is silver or silver splashed with white.) Due to the extended black base, blue-colored ducks are affected by juvenile white and aging white in a manner similar to blacks.

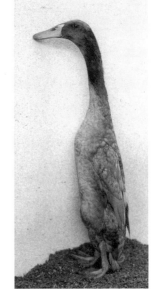

Breeds
East Indie, Call, Runner, Magpie, Ancona, Crested, Cayuga, Swedish

Genotype
E/E, Bl/bl

Day-Old Description
Ducklings have bluish-gray down, dark blue or green bills, and gray legs and feet shaded with yellow.

Adult Description
Blue plumage color is highly variable, even among full siblings. Ideally, it is a medium to medium-dark shade of rich blue gray with as little foreign color as possible. Drakes typically

A prize-winning Cumberland Blue Runner drake.

are darker than ducks. The head and neck are darker than the rest of the plumage. Bills are green blue in drakes and nearly black in ducks. Feet and legs are gray with more or less orange shading.

The plumage is enhanced when the feathers are laced with dark blue. Lacing is most pronounced over the shoulders and on the back, and drakes normally are the most distinctly laced. Lacing fades as the feathers age.

Blue dilution "leaks," allowing black to show through in the form of small flecks on individual feathers, entire feathers, or patches of feathers (rarely, an entire wing or side of a bird is black).

Common Faults

For exhibition, avoid birds with pronounced brown or pale blue color and conspicuous white feathering. Blues often develop white mottling as they age. As blue feathers are exposed to sunlight, they often develop a brown or yellow hue. Sound color is restored when old feathers are molted and new plumage is grown. An inadequate diet can also cause poor color.

Breeding Hints

Blue drakes mated to blue ducks produce offspring in a ratio of 1 Black: 2 Blues: 1 Silver. Blue x Black matings produce half Blacks and half Blues, while Blue x Silver produce half Blues and half Silvers. Silver x Black produces all Blue offspring. Breeders often debate the pros and cons of these different matings, but all four of them can produce excellent blue specimens. The main advantage of the Blue x Blue mating is that birds with distinct lacing can be chosen for both sides of the mating. To avoid having blue ducks that gradually become pale and washed-out, it is helpful to select breeders with rich color.

Chocolate

Chocolate is a rare color in ducks of Mallard descent. Genetically, it is extended black plus brown dilution. Due to the extended black base, chocolate-colored ducks are affected by juvenile white and aging white in a manner similar to black ducks.

Breeds
Call, Runner, Magpie, Ancona

Genotype
Usually: E/E, d/d; occasionally: E/E, d/d, bu/*bu*

Day-Old Description
Ducklings are dark brown (often with a yellowish blush on the chest and around the bill) with blackish-brown bill, legs, and feet.

Adult Description
As noted under genotype, there are two forms of Chocolate in ducks of Mallard descent. In the more common form, both genders are dark brown (sometimes appearing black from a distance), often with green sheen (especially on the head, neck, back, and rump of drakes), dark green to blackish-brown bills, and dark-brown legs and feet that are shaded with orange in mature drakes.

The second form of Chocolate is rare, not as dark in color, and normally lacks the green sheen.

A pair of day-old Chocolate Anconas showing the broken pattern of their down, feet, and bills.

Common Faults

For exhibition, avoid birds with brown lacing, penciling, or obvious white in their plumage. This color fades under prolonged exposure to sunlight, so be careful about culling Chocolates for pale or uneven color.

Breeding Hints

Birds that do not develop aging white in their plumage until they are older make valuable breeders. Type can be improved by out-crossing onto Blacks possessing outstanding conformation. The mating of Black drakes onto Chocolate ducks produces all Black offspring, but the sons are Chocolate "carriers." If these "split" sons are mated back to Blacks, half of their daughters will be Chocolates, or, if they are mated to Chocolates, half of both their sons and daughters will be Chocolate. If a Chocolate drake is mated to Black ducks, it is a sex-linked mating and all sons will be Blacks that carry one Chocolate gene, and all daughters will be Chocolate.

White

White is the most common color in domestic ducks. Genetically, white ducks are colored, but two inhibitor genes prevent the production of feather pigment. (To find out what colors lie hidden "under" the white, cross a White onto a purebred Gray. Any color characteristics in the first-generation offspring other than wild type were carried by the White parent. To find possible recessive color genes, mate the F_1 offspring together and hatch as many F_2 offspring as possible. You may uncover a new color!)

Breeds

Call, Mallard, Campbell, Runner, Crested, Aylesbury, Pekin

Genotype

c/c

Day-Old Description

Ducklings have yellow down and bills and orange legs and feet. Pink-billed breeds such as the Aylesbury have pink bills at hatching. A few strains produce some offspring with more or less gray or brown in their down. These birds normally are pure white at sexual maturity.

Adult Description

Depending on how much yellow and orange pigment these ducks consume, their plumage ranges from light canary yellow to silky white. Bills are yellow to orange, while the legs and feet are bright orange. Eyes are grayish blue. As females reach sexual maturity, it is common for more or less black or green spotting to develop in their bills due to hormonal changes.

Common Faults

One reason this variety is so popular is they have few color faults. There is virtually no challenge to breeding ducks with good white plumage.

For exhibition, creamy yellow plumage is preferred in Pekins, whereas satin white is desirable in all other breeds. Breed standards call for clear-yellow or orange-yellow bill color and clear-orange leg and foot color. Black or green spotting in the bills of drakes is a disqualification, while it is only a point cut in ducks. In general, ducks with dark color in their bills are more productive layers than their clean-billed strain mates.

Breeding Hints

Normally, the only characteristic that breeders of Whites need to pay attention to is the color of the bill. If you want good productivity, be careful about putting excessive emphasis on clean bill color in sexually mature ducks.

Bibbed Pattern

The white bibbed pattern found in standard varieties of ducks in North America is probably the result of a pair of dominant bib genes combined with extended black. The size and shape of bibs vary considerably, even among siblings.

(*Note:* Dominant bib appears to be linked with both extended black and dusky, which explains why solid varieties carrying E or M^D produce a percentage of offspring possessing white on the neck front, even if the breeding stock is carefully selected for no white. A rare recessive white bib also exists which has occasionally been found in mallard-colored ducks. No standard varieties in North America carry recessive bib.)

Breeds

Call, Swedish, rarely Runner

Genotype
White-bibbed Blue: Bl/bl, s/s
White-bibbed Black: bl/bl, s/s

Day-Old Description
In White-bibbed Blues, the down is blue gray, ideally with yellow only on the chest and on the wing tips. However, many ducklings have more or less yellow on the face, especially around the bill and behind the eyes. Bills vary considerably, but often are dark green or blue marked with more or less yellow. Legs and feet are dark gray shaded with more or less yellow.

Adult Description
On average, drakes have larger bibs than ducks. Ideally, bibs are solid white, symmetrically shaped, and have clearly defined borders. The exact size is not critical, but bibs that spill onto the underbody are overdone, whereas a mere spot is too small. Breed standards call for the outer two or three flight feathers to be white. The exact number of white flight feathers is a relatively minor detail that should not be overemphasized by breeders or judges. Drakes typically have green to greenish-blue bills while those of ducks are dark blue. The legs and feet are orange shaded with gray (nearly black in some females).

Common Faults
For exhibition, avoid birds with considerable white in their faces, bibs that are broken with color, and blotchy colored bills (bill color often, though not always, darkens with age). It is common for bibbed ducks (especially females) to develop white mottling in the colored portion of their plumage as they age.

Breeding Hints
Perfecting the bibbed pattern is a challenge. Choose breeders with the cleanest and most symmetrical bibs possible. Keep in mind that, usually, drakes with small bibs will produce daughters with even smaller bibs and ducks with large bibs will produce sons with huge bibs and white underbodies. Generally, the best results are obtained when drakes with moderately large bibs are mated to ducks with medium-sized bibs. The more ducklings that are hatched, the better the chances are of producing some birds with outstanding markings. With practice, the best-marked birds can be identified the day they hatch (including how many white flights they will have).

Runner Pattern

The white pattern on a colored background in Penciled and Fawn & White ducks is the result of a pair of runner-pattern genes.

Breeds
Call, Runner

Genotype
Penciled: M^D/M^D, d/d, R/R
Fawn & White: M^D/M^D, Bl/Bl, d/d, R/R

Day-Old Description
The down pattern of day-olds is a replica of the adult pattern. Penciled ducklings are brown and yellow. In Fawn & Whites, the fawn portions of the down are only slightly darker than the yellow portions.

A champion Penciled Runner drake with outstanding markings.

Adult Description
Ideally, the white areas include the upper two-thirds of the neck, the throat, a finger that extends from the back of the head to and partially encircles the eye, a line that divides the bill from the cheek patches, a belt across the underbody and the outer two-thirds to three-quarters of the wings.

Penciled ducks are genetically khaki-colored with the addition of the runner pattern. Fawn and white ducks are penciled birds with two blue-dilution genes.

Common Faults
The runner pattern (along with the magpie) is the most challenging to perfect. Therefore, breeders and judges should not overemphasize minor details, but rather focus on overall clarity and symmetry. Common faults include excessively large or small head caps and cheek markings, head and cheek markings that trail down onto the neck, colored patches on the throat and neck, "snow" flecks on the shoulders and back, lack of white belt across the underbody, colored wing feathers, and solid-white tail feathers.

Breeding Hints

From one generation to the next, white tends to increase in drakes and color increases in ducks. Drakes with small cheek patches tend to produce daughters with medium-sized cheek patches, whereas ducks with large cheek patches tend to produce sons with medium-sized patches. When possible, use breeders with clearly defined markings. However, poorly marked birds that are out of a good strain can produce some excellent marked offspring.

The more ducklings hatched, the better the chances of some birds with outstanding markings. With practice, the best-marked individuals can be identified the day they hatch. Breeders who produce well-marked runner-pattern ducks have perfected one of the two most challenging colors in ducks.

Magpie Pattern

The bold magpie pattern results from combining a large dominant bib, runner pattern, and extended black. This genotype is highly variable in its expression, resulting in wide variation in the plumage pattern of even the most carefully bred strains of magpie-marked ducks.

Breeds
Call, Magpie

Genotype
Black Magpie: E/E, S/S, R/R
Blue Magpie: E/E, Bl/bl, S/S, R/R
Chocolate Magpie: E/E, d/d, S/S, R/R

Day-Old Description
The down pattern of the day-olds is a replica of the adult pattern. Their bills, legs, and feet are yellow, sometimes with black or gray markings. (Some ducklings hatch out solid yellow and will be white at maturity.)

Adult Description
Ideally, the crown of the head and the back mantle (including the shoulders, back, and tail) are colored, with the rest of the plumage being white. The bills are yellow or orange with green shading in young birds, gradually darkening with age. The legs and feet are orange red shaded with grayish black, darkening with age, especially in the females.

Common Faults

The magpie pattern (along with the runner pattern) is the most challenging pattern to perfect in ducks. Therefore, both breeders and judges should not place undue emphasis on minor details, but rather emphasize overall clarity and symmetry of the pattern. Common faults include head caps that cover less than half of the crown or are totally missing (the colored caps gradually disappear on most females as they age), caps running down back of the neck or spilling below the eyes, color on the breast or sides of body, and white on the shoulders, back, and tail. With age, the colored portions of the plumage will often gradually turn white, especially in females.

Breeding Hints

As mentioned above, white tends to increase in drakes and color increases in ducks from one generation to the next. When a standard-marked drake is mated to a standard-marked duck, they tend to produce excessively white sons but well-marked daughters (and a fair number of solid-white offspring). In general, females with excessive color (including color under the eyes) produce the best-marked sons.

If they are descended from a good strain, even poorly marked Magpies can produce some offspring with excellent markings. The more ducklings that are hatched, the better the chances of producing some outstandingly marked birds. With practice, the best-marked birds can be identified the day of hatching. Like breeders of runner-pattern ducks, breeders who produces well-marked magpie-pattern ducks should be satisfied knowing they have perfected one of the top two most challenging duck colors.

Ancona (Broken) Pattern

The ancona pattern results from combining a small- to medium-sized dominant bib, runner pattern, and extended black. Selective breeding has produced the crazy quilt effect of this pattern. The wildest marked individuals normally carry a single dose of extended black, whereas magpie-pattern varieties carry a double dose of E.

Breeds

Call, Ancona

Genotype
Black Ancona: E/e, S/S, R/R
Blue Ancona: E/e, Bl/bl, S/S, R/R
Chocolate Ancona: E/e, b/b, S/S, R/R
Lavender Ancona: E/e, Bl/Bl, b/b, S/S, R/R

Day-Old Description
The pattern of the day-olds is a replica of the adult plumage. (Some duck-lings hatch out solid yellow and will be white at maturity.) The bill, legs, and feet are yellow or orange, usually marked with more or less black or dark brown.

Adult Description
There is no set pattern. In fact, the more haphazard and asymmetrical the markings, the better. Ideally, there should be bold patches of color under the eyes, on the chest, on the sides of the body, and on the back. A dramatic pattern is paramount. Because the best-marked individuals carry a single extended black gene, blacks and blues (especially the drakes) often have reddish-brown shading in the colored portions of their plumage. This "rust" is a fault in all other patterns, but is a characteristic of the broken pattern. The bill, legs, and feet should be as spotted and blotchy as possible in young mature birds, with these extremities turning darker and often more solid with age.

Common Faults
The most common pattern faults include no color patches below the eyes or on the chest, sides of body, or back.

Breeding Hints
To reduce the number of solid-white offspring produced, avoid mating together two birds that have less than one-third of their plumage colored. Ancona-pattern ducks may produce some offspring with runner or magpie patterns. If these are mated back to broken-pattern birds, they will produce some offspring with good broken pattern.

Muscovy Colors

The APA recognizes only four color varieties of Muscovies: White, Black, Blue, and Chocolate. However, more than two dozen varieties are raised in North America. These colors and patterns are the result of various combinations of wild-type genes plus one or more of the nine main mutations as outlined in the table below.

Plumage Color and Pattern Genes in Muscovy Ducks

Name	Symbol	Relationship to "Wild"	Main Visual Effects on Adults
Wild Pattern	A^+	Wild type	Wild Muscovy pattern
Atipico Pattern	a	Recessive	Extends dark pigment through duckling; reduces brown markings in adult
Non-blue	n^+	Wild type	Permits normal plumage color expression
Blue	N	Incompletely dominant	Dilutes black pigment to blue gray in single dose, to silver in double dose
Non-chocolate	Ch^+	Wild type	Permits normal plumage color expression
Chocolate	ch	Sex-linked recessive	Changes black pigment to reddish brown
Non-lavender	L^+	Wild type	Permits normal plumage color expression
Lavender	l	Recessive	Dilutes black pigment to uniform lavender
White	P	Incompletely dominant	Prevents pigmentation of plumage
Colored	p^+	Wild type	Permits normal plumage color expression
Non-barred	B^+	Wild type	Permits normal plumage color expression
Barred	b	Recessive	Lightens duckling color; causes more or less barring in juvenile feathers
Non-rippled	Br^+	Wild type	Permits normal plumage color expression
Brown-rippled	br	Recessive	Lightens ground color; causes irregular crosshatches on feathers
Non-pied	D^+	Wild type	Permits normal pigmentation throughout
Duclair Pied	d	Recessive	Inhibits pigmentation on neck, wings, and underbody
Canizie	C	Dominant	Inhibits pigmentation on head and neck
Colored Head	c^+	Wild type	Permits normal pigmentation throughout plumage

Wild Type

This is the original color found in wild Muscovies. All other varieties are mutations. The wild color is sometimes seen in feral populations as well as in some domestic flocks.

Genotype
A^+/A^+, n^+/n^+, Ch^+/Ch^+, L^+/L^+, p^+/p^+, B^+/B^+, Br^+/Br^+, D^+/D^+, c^+/c^+

Day-Old Description
Ducklings have the classic dark and yellow camouflage pattern. Compared with Mallards, the light portions of the down are paler yellow, the dark facial stripe is abbreviated and does not run between the eyes, and the bill and the dark portions of the head and neck have a browner cast. The bill is black and the legs and feet are dusky yellow with black shading.

Adult Description
The plumage is black with considerable green and purple iridescence. The breast, sides, and underbody are sometimes marked with more or less brown or dark bronze. (The feathers of the juvenile plumage are typically marked brown.) The wings start out solid colored, with the forewings gradually turning white over the course of several years. The facial skin patch is normally smooth and pigmented with considerable black.

Common Faults
There are few domestic Muscovies with good wild-type color. The most common faults are the presence of white in the body plumage and the lack of black pigment in the facial skin patch.

Breeding Hints
There is no written standard for the wild-type color. However, using wild birds as the standard, select for breeding those specimens that: have the classic dark and yellow camouflage pattern as ducklings; display more or less brown or bronze shading in their juvenile plumage; and as adults exhibit brilliant green and purple iridescence on the surface of their feathers and have as much black as possible on their facial skin patches.

Black

From 1904 until 1998, this variety was officially known as Colored. Genetically, Blacks are wild type except for two atipico genes. (The atipico gene in Muscovies functions in a fashion similar to the dusky gene in Mallard-derived breeds.)

Most Muscovies carrying the atipico gene will have a small white patch at the juncture of the bill and throat.

Genotype
Wild type except for a/a

Day-Old Description
Ducklings are solid black (sometimes with a spot of yellow on the front of the neck) with brown shading especially on the head and neck. The bill, legs, and feet are black to dusky black.

Adult Description
The plumage is black with green and purple iridescence. (The iridescence is less pronounced in the juvenile plumage.) The wings usually start out solid colored, with the forewings gradually turning white over the course of several years. With age, more or less white flecking often develops in the head and neck plumage.

Common Faults
For exhibition, faults to avoid include: pronounced black in the facial caruncles (even the best drakes normally have a bit of black near the eyes), excessive white in the plumage other than the forewings, lack of white forewings in mature specimens, and pronounced brown in the plumage of mature birds (most mature "Blacks" with brown in their plumage are actually wild type). A white spot on the throat at the juncture of the bill is not a fault.

Breeding Hints
In general, individuals with the blackest feet and legs have more black or mulberry in their faces. Therefore, it is usually helpful to select breeders with dusky yellow feet and legs. Because the white flecking of the head and neck plumage often does not develop until a bird is 6 to 18 months old, 2-year-old birds with the least white in these areas make valuable breeders.

Blue

Genetically, these are wild type or Black with one blue-dilution gene — the latter preferred for exhibition. (Two blue genes result in silver-colored offspring.)

Genotype
Wild type except for N/n^+ or a/a, N/n^+

Day-Old Description
In Blue ducklings, the black portions of the down are diluted to bluish-gray and the bill, legs, and feet are slightly lighter than those of Blacks.

Adult Description
The plumage is bluish gray, ideally with each feather of the back and sides of the body laced with a darker border. Often, drakes are darker colored with more pronounced lacing than ducks. (The juvenile plumage is duller.) Everything else is similar to Blacks.

Common Faults
Even full siblings can vary considerably in the shade of their plumage. The exact shade of blue gray is not critical, although medium to medium-dark birds are generally preferred for exhibition. Black tends to "leak" through to produce "ink spots" that range in size from a small streak to patches of solid-black feathers. Blues can have the same faults as Blacks.

Breeding Hints
Excellent-colored Blues can be produced by three matings. Blue x Blue produces ducklings in a ratio of 1 Black: 2 Blues: 1 Silver. Blue x Black yields half Blues and half Blacks. Silver x Black gives all Blue offspring.

Chocolate

Genetically, these are wild type or Black plus chocolate; the latter is preferred for exhibition. Chocolate is sex-linked.

Genotype
Wild type except for ch/*ch* or a/a, ch/*ch*

Day-Old Description

Chocolate ducklings are a rich reddish brown with dark-brown bills, legs, and feet.

Adult Description

The plumage is a rich reddish brown with purple, and in the right light a greenish iridescence is detectable. (The juvenile plumage is duller.) The bill is pink (sometimes with brown shading) and the feet and legs are yellowish brown. The white markings are the same as in Blacks. Because black pigment is diluted to brown, Chocolates do not have black in their faces.

Common Faults

Chocolates often display faded plumage. This is usually the result of exposure to sunlight. A well-balanced diet and shade help Chocolates stay in good feather condition longer.

Breeding Hints

Type and size can be improved by mating Chocolates to outstanding Blacks. A Chocolate drake mated to a Black duck produces all Black sons (that are Chocolate carriers) and Chocolate daughters. The reciprocal cross of Black drake to Chocolate duck produces all Black offspring, with the drakes being Chocolate carriers.

White

Unlike in Mallard-derivative breeds, white plumage in Muscovies is caused by a pair of dominant genes. Muscovies that carry a single white gene are haphazardly bicolored.

Genotype

P/P

Day-Old Description

Ducklings are lemon yellow with pink bills and yellow legs and feet. Sometimes they have a colored spot on the top of their heads.

Adult Description

Most adults are pure white, even if they possessed a colored head spot in their juvenile feathering. The bills are pink to pinkish white, and the legs and feet are yellow to orange.

Common Faults
In first-year birds, color on top of the head is not a disqualification for exhibition. White Muscovy drakes are prone to developing dark brown or black markings in their bills, which is a show disqualification.

Breeding Hints
For the breeder of exhibition Whites, it can be useful to choose clean-billed breeding drakes that are at least 2 years old and mate them to females with no black or brown in their bills.

Nonstandard Varieties

The following varieties have not been recognized by the APA and have no written standards. All of them are attractive and several are common in some parts of the world.

Lavender

This lovely, rare color is sometimes called Self Blue since each feather is uniformly colored throughout its surface. To confuse matters further, some people call this variety Silver, despite the fact that the phenotype and genotype of Lavender and Silver are distinctly different. True Silvers are lighter in color (and lack the purplish hue of Lavenders) and genotypically are black with two blue dilution genes. On the other hand, Lavenders are black with two lavender genes.

Genotype
Wild type except for l/l or a/a, l/l

Day-Old Description
Lavender ducklings are medium bluish silver, a shade or two darker than true Silvers (N/N).

Adult Description
The plumage is a uniform bluish lavender without darker lacing on the

perimeter of the feathers or black ink spots. The white markings are the same as in Blacks. The bill is pinkish, often with a dark saddle, and the legs and feet are gray.

Common Faults
Lavenders can have the same faults as Blacks. This color will eventually fade and discolor when exposed to bright sunlight. The true color is restored following the molt.

Breeding Hints
The best-colored Lavenders have a black genetic base rather than wild type. To improve the size and conformation of Lavenders, outcross them onto the best Blacks obtainable. The first generation offspring will (under normal circumstances) be black, but when the F_1s are intermated, they will produce F_2 offspring in a ratio of 3 Blacks: 1 Lavender.

Barred

Barred Muscovies are regionally common. Genetically, these carry two barring genes. Barred Muscovies can be produced in Black, Blue, Chocolate, and Lavender.

Genotype
b/b

Day-Old Description
The down of Barred Muscovy ducklings resembles that of Harlequins with their yellow down tipped with color. The bills, legs, and feet are a shade or two lighter than in nonbarred individuals.

Adult Description
Barring is fairly distinct in most juveniles, but largely disappears in the adult plumage where it sometimes causes a "marbled" effect, especially on the sides and underbody. With careful selection of breeding stock, barring can be enhanced in the adult plumage. Mature birds have white forewings. In some countries, the barred color is preferred for commercial production due to the lighter undercolor it generates.

Common Faults
Barreds can have the same faults as Blacks.

Breeding Hints
Size and type can be improved by outcrossing Barreds to the best Blacks, Blues, Chocolates, or Lavenders available. The F_1 generation will be non-barred, but one out of four of the F_2 offspring will be barred. The barred offspring are easily identified at hatching by their nearly yellow down.

Brown-Rippled

This distinctive color is rare in most localities. People sometimes mistakenly call it Barred, but the color of the hatchlings is diagnostic. Genetically, these are wild type or black with two brown-rippled genes. (Rippled can also be combined with blue, chocolate, or lavender.)

Genotype
br/br

Day-Old Description
At first glance, these can be mistaken for Blacks. However, upon closer examination, it is evident that the body, bill, legs, and feet are browner than in Blacks.

A distinctly marked Brown-rippled old Muscovy drake (bred on our Waterfowl Farm).

Adult Description

The base color is blue gray, irregularly marked with darker cross-hatches that vary from dull black to dark brown. Some birds have vertical center-marks on feathers (especially over the shoulders and back). Upon seeing this variety for the first time, people often assume they are mismarked Blues. They have the white forewings as in Blacks.

Common Faults

Brown-rippled Muscovies can have the same faults as Blacks.

Breeding Hints

Size and type can be improved by using the same methods as with Barreds. The rippled F_2 offspring will be slightly lighter in down color than their non-rippled siblings.

White-Heads

This unique pattern is common in some feral populations and farm flocks. Genetically, these carry the canizie gene. White-heads are bred in Black, Blue, Chocolate, and Lavender.

Genotype

C/C

Description

In juveniles, the head and neck are colored, gradually turning white with age. In adults, the head and neck are white (often peppered with dark feathers), with the rest of the body colored like other Muscovies. Especially in some drakes, the white may spill down onto the chest and underbody.

Common Faults

Keep in mind that it can take a year or more for the white head to fully develop, and ducks tend to have better markings than drakes. White in the body plumage is a common fault.

Breeding Hints

Females that retain colored flecking in their head and neck plumage often produce sons with the best white-head markings. On the other hand,

Females with pure white heads and upper necks usually produce the best marked daughters.

Duclair (Magpie)

Muscovies with bold white markings are common in many countries. Birds of this genotype go by various names such as Duclair, Piebald, Colored and White, Parti-colored, and Magpie. Genetically, these carry two duclair genes. The duclair pattern can be bred into Muscovies of any color.

Genotype
d/d

Description
The duclair gene is highly variable in its expression. Some birds with this genotype will have considerably more white than others. In general, Duclairs have well-defined colored and white sections on their body plumage. (Muscovies that are haphazardly splashed or mottled with white usually are not Duclairs, but rather, carry a single gene for white C/c.) Some Duclairs have good magpie markings with a colored mantle that includes the shoulders, back, and tail, and they may or may not have color on their heads.

Common Faults
In Great Britain, magpie-patterned Duclairs are standard varieties and are called Black and White, Blue and White, and so forth. Common faults of these are poorly defined shoulder and back mantles, color spilling down on the sides of the body, and excessive color on the head.

Breeding Hints
By carefully selecting breeders for several generations that possess the preferred markings, pattern uniformity can be improved in the offspring.

Other Varieties

Each of the following rare varieties results from combining two or more of the colors previously described.

Buff

These are the result of combining chocolate with two brown-rippled genes (ch/*ch*, br/br). Adults are a medium shade of buff brown, with white forewings.

Blue Fawn

These are the result of combining chocolate with one blue-dilution gene (ch/*ch*, N/n$^+$). Adults are blue brown, with white forewings.

Lilac

These are the result of combining chocolate with two blue-dilution genes (ch/*ch*, N/N). Adults are pale lilac with white forewings.

Pastel

These are the result of combining chocolate with two lavender genes (ch/*ch*, l/l). The adults resemble Lilacs, but are richer colored.

Loonie

These are the result of combining two barred with two brown-rippled genes (b/b, br/br). I have not bred these, but the fine geneticist W. F. Hollander produced and named these, describing them as resembling the color of the Common Loon.

ELEVEN

ACQUIRING STOCK

Having selected the breed or breeds you want to raise, the next step is locating suitable stock. The importance of starting with good-quality birds cannot be overemphasized. The productivity, growth rate, and size of ducks within the same breed vary a good deal. If your duck project is going to be economically practical and free of unnecessary problems, healthy and productive birds are essential.

Production-Bred Stock vs. Standard-Bred Stock

Egg- and meat-producing characteristics are given first priority in production-bred ducks, with less concern for perfect color or shape. On the other hand, standard-bred birds are used for showing in competition and are painstakingly selected for color, size, and shape, with production abilities often given second priority. In some breeds of ducks, the productivity and growth rate of exhibition strains are equal to or better than the commercial stock that is commonly sold by hatcheries. Normally, standard-bred stock is priced higher than production-bred stock.

Hatching Eggs

If a dependable setting hen or incubator is available, you may wish to buy hatching eggs to start your flock. Some advantages of this method are that hatching eggs normally sell for one-third to one-half of the prices of day-old ducklings, and you get to experience the fun of waiting for and witnessing the hatch. Some disadvantages are that eggs vary in their fertility, they may be broken or internally damaged when shipped, and it is impossible to know just how many ducklings will hatch.

If you receive a shipment of hatching eggs that is insured or you are paying COD, open the package in the presence of your postal carrier to check for breakage and to count the number of eggs received. If a substantial number of eggs are broken or there are fewer eggs than you paid for, the postal carrier will provide a claim report.

Unless you know that the eggs are more than 2 weeks old when you receive them, higher hatchability can often be obtained if shipped eggs are held at 55 to 65°F 12 hours prior to being placed in the incubator. Older eggs are best set promptly upon their arrival.

Day-Old Ducklings

Purchasing day-old ducklings is the most popular method of starting a duck flock. They are more widely available than hatching eggs or adult stock. Ducklings are sturdy and can be shipped thousands of miles successfully. When ducklings hatch, they have approximately two-thirds of the yolk left and do not need to eat or drink for several days while they are living off the yolk, making this the best time to ship a bird.

Ducklings are sold sexed or straight-run (the sex ratio that nature provides). Theoretically, unsexed ducklings run 50 percent drakelets and 50 percent ducklets. Practically, there may be considerably more males than females or vice versa, particularly when ducklings are purchased in small quantities.

When ordering ducklings, give your telephone number and instruct the shipper to include it on the shipping label. If you live on a long rural route, ask your local postmaster to hold the ducklings at the post office and phone you upon their arrival so you can pick them up promptly. When a shipment is received, open the box in the presence of the postal employee, check the condition of the ducklings, and count the live birds. If you receive fewer live ducklings than you paid for, the postal carrier should provide a claim report.

Care of Shipped Ducklings

The first 24 hours after ducklings arrive are critical. The birds should be given drinking water and appropriate food, and be allowed to rest in a pre-heated brooding area as soon as possible.

As you take the little ones from the shipping box and place them in the brooder, dip each of their bills in lukewarm water (to which 1 teaspoon of honey or corn syrup per quart of water has been added). Sprinkle finely chopped lettuce, dandelion greens, or tender, young grass on the water. Let them drink and eat greens for 15 to 30 minutes prior to giving them other food.

Ducklings should be checked frequently the first day, but do not handle or disturb them more than absolutely necessary. Always use waterers that ducklings can drink from easily but cannot get into and get soaked.

Buying Mature Stock

The quickest way to obtain a producing duck flock is to purchase mature birds. Poultry farms and hobbyists sometimes have adult stock available. Waterfowl adapt quickly to new climates and are readily shipped, so you can order from out-of-area breeders if the birds you want are not available locally.

Some good places to look for duck sources include: feed stores; agriculture fairs; agriculture extension services; classified ad sections of poultry, farm, and gardening magazines and local newspapers; and appendix F (Duck Breeders and Hatchery Guide) in this book.

Day-old Khaki Campbell, Mallard, and White Indian Runner ducklings several hours after being taken out of the incubator.

How Many?

The ideal number of ducks depends on your purpose for raising them, the breed raised, environmental conditions, and your management.

To estimate the number of ducks needed for a laying flock, calculate the total number of eggs desired over a year. Divide this number by the average number of eggs a duck of the breed you are going to raise will lay yearly, and then add 10 percent to allow for the occasional poor layer or mortality. Keep in mind that the figures in the Breed Profile Chart (see chapter 3, page 20) for the yearly egg production of the various breeds are for ducks that are fed concentrated feeds and exposed to no less than 14 hours of light daily during the laying season.

When purchasing breeding stock as day-olds, plan on culling out 10 to 50 percent of the inferior birds at maturity. Remember, no matter how carefully bred, not all offspring are suitable for breeding purposes.

When purchasing show birds as day-olds, plan on raising a minimum of two or three ducklings for every show bird desired. In general, the more ducklings raised, the better your chances of raising an elite show bird.

Gotcha!

When Mr. DeFord told me I could have any of the ducks I caught in his gravel pit, he must have thought the chances were slim that a 12-year-old could capture those half-wild Mallards. The first step in my plan was to shamelessly bribe the ducks with wheat every day after school. I then constructed a C-shaped trap from 4-foot-high chicken wire and metal fence posts pounded into the ground. After a week of baiting, I put down the feed in the back of the enclosure and hid behind a nearby log. The ducks rushed in to grab the feed as soon as I was out of sight. Moving quickly, I dashed to the front of the trap and closed it with an attached flap of wire. Within an hour, the flock of surprised broadbills had a new home in my duckyards.

TWELVE

INCUBATION

For many of us who own poultry, the incubation and hatching of eggs is the most fascinating phase of raising birds. When holding an egg in one's hand, it is difficult to comprehend that inside the shell there exists every element necessary for the beginning and growth of a new life. In fact, when a fertile egg is laid, an embryo several thousand cells in size has already formed. If stimulated by warmth and movement, that tiny spark of life will grow, break from the shell, and present itself as part of a new generation of ducks!

Hatching Eggs

Any egg that is fertile has the potential to hatch. However, for consistently good results, hatching eggs need to be produced by healthy ducks that live in a good environment, are not obese, and consume an adequate diet. (In general, breeding feeds have higher levels of protein and most vitamins when compared to layer rations. If you cannot find a duck or waterfowl breeder feed in your local feed stores, a gamebird breeder feed will normally work well.)

Once an egg that is going to be hatched is laid, the use of proper handling, storing, and incubation procedures will increase its chances of producing a viable duckling.

Gathering

Eggs that are going to be incubated by a foster hen or in an incubator should be gathered at least twice daily to protect them from predators and prolonged exposure to the elements. Hatching eggs must *always* be handled gently so that the diminutive embryo is not injured or the protective shell cracked. Do not roll eggs over and over, jolt them sharply, or handle them with dirty hands — all of these can destroy fertility.

Cleaning

Eggs that are nest-clean hatch the best. Dirty eggs need to be washed as soon after gathering as possible.

Washing does have several negative effects on duck eggs. It causes removal of the cuticle (a protective film on the shell that reduces dehydration and screens out pollutants), resulting in slightly lower hatchability and the need to raise the humidity level by 5 to 10 percent during incubation. Nonetheless, it is preferable to wash dirty eggs rather than set them uncleaned, since contaminated eggs create an unsanitary condition under the hen or in the incubator, and frequently explode during incubation due to the buildup of pressure caused by harmful gases within the shell.

When eggs are washed, always use clean water that is 10 to 25°F warmer than the eggs. Washing with dirty water spreads contaminants from egg to egg, while washing with cold water forces filth deeper into the shell pores. In situations where it is practical, use a hatching-egg disinfectant in the wash water.

Selecting

Not all fertile eggs are suitable for hatching. Those used for setting should have normal shells and be average to large in size. Extremely large eggs often have double yolks and seldom hatch. Eggs having irregular characteristics or cracks are best used for food. Valuable eggs with small cracks can sometimes be saved by placing a piece of masking tape over the fracture.

Storing

Proper care of eggs *prior to setting* is just as important as correct incubation procedures. In working with owners of small poultry flocks, I have found that careless handling of eggs before incubation is one of the leading causes of bad hatches. Always keep in mind that no matter how faithful a setting hen is or how carefully the incubator is regulated, a poor hatch will result if the embryos have been weakened or killed during the holding period.

Where

Eggs must be stored away from direct sunlight in a cool, humid location. Cellars and basements are ideal places; refrigerators are usually too cold.

Position

The position in which eggs are stored prior to incubation has little effect on hatchability. The results of a study involving thousands of duck and goose eggs showed no significant difference in the hatchability of eggs stored on their sides, vertical with the air cell up, or vertical with the air cell down. However, when eggs are stored in 12-egg cartons or 30-egg flats (20-egg turkey flats are best for large duck eggs), less breakage occurs if the eggs are positioned with their large end up.

Temperature

The ideal storage temperature for hatching eggs that are held for 10 days or less seems to be 55 to 65°F. If eggs are kept for a longer time, a temperature of 48 to 52°F will produce better hatches. Because wide temperature fluctuations reduce the vitality of embryos, it is wise to store eggs where the temperature stays at a fairly constant level. In a study of the effects of storage temperature on hatchability in duck eggs that were stored 10 to 14 days, the following results were obtained:

STORAGE TEMPERATURE	HATCHABILITY PERCENTAGE
38°–40°F	61
60°–62°F	73
76°–82°F	42

Turning during Storage

Duck eggs that are held 5 days or less show little improvement in hatchability when turned during the storage period. On the other hand, when eggs are stored longer than 5 days, the hatch can be increased 3 to 15 percent by turning them daily while they're being saved for incubation. If eggs are stored in egg cartons or flats, they can be turned by leaning one end of the container against a wall or on a block at an angle of 30 to 40 degrees each day, alternating the end that is raised.

Length of Storage

Ordinarily, the shorter the storage period, the better the hatch. A few eggs that have been held 4 weeks or longer may hatch, but for good results the general rule is not to keep eggs for more than 10 days before setting them. The negative effect of long storage periods on hatchability can be seen in the results obtained from a test involving several thousand eggs.

LENGTH OF STORAGE	HATCHABILITY PERCENTAGE
1–7 days	71
8–14 days	64
15–21 days	47
22–28 days	18

Incubation Period

The normal incubation period for Mallards and their derivatives is 26 to 29 days, while Muscovies require approximately a week longer — 33 to 35 days. High temperatures during storage or incubation cause premature hatches, while long storage periods and low incubation temperatures result in late hatches.

Average Fertility

It is unusual for all eggs in a large setting to be fertile. The average fertility for heavyweight breeds is 85 to 95 percent and for lightweight birds, 90 to 98 percent. See the chart on Pinpointing Incubation Problems, page164, for common causes of poor fertility.

Average Hatchability

The hatchability of artificially incubated duck eggs often is 5 to 10 percent lower than that of chicken eggs. Still, good setting hens frequently hatch every fertile duck egg they incubate. Under artificial incubation, the average hatchability falls between 65 to 85 percent of all eggs set, or 75 to 95 percent of the fertile eggs. See the chart on page 164 for common causes of poor hatchability.

Natural Incubation

When you wish to hatch a moderate number of ducklings, natural incubation is often the most practical. A good setting hen is a master at supplying the precise temperature, and instinctively knows just how often eggs need to be turned. She also serves as a ready-made brooder, eliminating the need to supply an artificial source of heat.

Choosing Natural Mothers

In the Breed Profile Chart in chapter 3 (see page 20), the section on Mothering Ability indicates the average capability of the various breeds as natural mothers. Duck eggs can also be hatched by goose, turkey, and chicken hens. Some of the breeds of chickens that make the best foster mothers are Silkie or common barnyard bantams and large Old English Games, Orpingtons, and Cochins.

Clutch Size

Duck hens normally cover 8 to 14 (Muscovies 16 to 20) of their own eggs. Some hens lay such large clutches that they cannot incubate the eggs properly. In this situation, the oldest eggs — those that are the dirtiest — should be removed, leaving only the number that the hen can cover comfortably. Eggs must be positioned in a single layer to hatch well, never stacked on top of one another. If too many eggs are in a nest, the result will be a poor hatch or a complete loss.

Care of the Broody Hen

Setting hens are temperamental and should not be disturbed by people or animals. It is advantageous to isolate the broody from the rest of the flock with a temporary partition. This precaution will keep other hens from disrupting the incubation process by attempting to lay in the broody's nest. Unlike chickens, duck hens and their nests usually *cannot* be moved.

To remain healthy during her long vigil on the nest, the hen must eat a balanced diet, have clean drinking water, and be protected from the hot sun. Feed and water containers should be placed several feet from the nest so that the hen must get off to eat and drink. A leave of absence from the nest for 5 to 30 minutes once or twice daily is essential to the hen's good health and will not harm the eggs.

When chicken or turkey hens are used to hatch duck eggs, they should be treated for lice and mites several days before their setting chores commence. It is usually necessary to sprinkle waterfowl eggs with lukewarm water several times each week when turkey or chicken hens are employed.

Multiple Broods

Muscovies (and occasionally hens of other breeds) will frequently bring off two broods a year or up to three or four in mild and tropical climates. Multiple broods can be encouraged by feeding hens extra feed in the early spring and by removing the ducklings at 1 to 4 weeks of age.

Artificial Incubation

There are circumstances when an incubator is useful. Unlike setting hens, incubators can be used any season of the year, and come in such a wide range of sizes that any number of eggs, from one to thousands, can be set simultaneously or on alternate dates.

On the other hand, incubators must be attended to periodically each day to check the temperature and humidity, and eggs must be turned if this function is not performed automatically. The hatchability of eggs is lower and the number of crippled or weak young is higher under artificial incubation than with the natural method. Also, electric incubators are at the mercy of power failures unless a gasoline generator is available.

Types of Incubators

Incubators are made and sold in a wide range of sizes and shapes, with varying degrees of automation. They can be divided into two basic types: the still-air (gravity flow) and the forced-air.

Still-Air
These models, available with electric or oil heat, closely approximate natural conditions by placing the heat source above the single layer of eggs, causing

the upper surface of the eggs to be warmer than the lower portion. Still-air incubators are simple to operate, dependable, nearly maintenance-free, and are manufactured with capacities of 20 to 400 eggs. We have used four models of still-air machines and have had excellent results with each.

Forced-Air

These incubators are equipped with fans or beaters which move warmed air to all surfaces of the eggs, and normally have multiple layers of egg trays. Forced-air machines are available with capacities of 12 to many thousands of eggs. When compared with still-air incubators, they are better suited to automatic turning of eggs and take up less floor space for larger quantities of eggs. They are also more complicated, require greater maintenance, and sell for higher prices.

Homemade

With a little ingenuity and a lot of persevering care, satisfactory hatches can be obtained in a homemade incubator consisting of a cardboard or wooden box and light bulbs for heat. More elaborate incubators, complete with heating elements and thermostats, can also be crafted in the home shop. (County extension agents or 4-H officers often have plans available for building small incubators.) In emergency situations — such as a hen deserting her nest — it is possible to hatch eggs in an electric frying pan or on a heating pad.

Where to Place the Incubator

Incubators perform best in rooms or buildings where the temperature does not fluctuate more than 5°F over a 24-hour period. Consistent temperatures are especially important for still-air incubators, which should be located in a room with an average temperature of 65 to 75°F. *Do not* position your machine where it will be in direct sunlight or near a window, heater, or air conditioner.

Leveling the Incubator

Incubators, particularly still-air models, must be level to perform well. If the incubator is operated while askew, the temperature of the eggs will vary in different areas of the machine, causing eggs to hatch poorly and over an extended period of time.

Operating Specifications

Manufacturers of incubators include a manual of operating instructions with their machines. This guide should be carefully read and followed. The operating instructions often cannot be adapted from one machine to another with good results, particularly if one is a still-air model and the other a forced-air. If you acquire a used incubator that does not have an instruction booklet, manufacturers are usually willing to send a new manual if you send them a request with the model number of your machine.

Incubation Requirements

The following is a summary of the basic incubation requirements of duck eggs. Over time, you will fine-tune these procedures to meet the unique needs of your own situation and environment. Some of the factors that affect incubation requirements are climate (especially temperature, humidity, and barometric pressure), elevation, breeds being hatched, diet of the breeding stock, and the duration of the hatching season.

Setting the Eggs

Start the incubator *at least* 48 to 72 hours ahead of time and make all necessary adjustments of temperature, humidity, and ventilation before the eggs are set. People frequently put eggs in machines that are not properly regulated, thinking they can make fine adjustments after the eggs are in place. This practice is a serious mistake since one of the most critical periods for the developing embryo is the first 4 or 5 days of incubation.

Prior to being placed in the incubator, duck eggs need to be warmed up for 5 or 6 hours at a room temperature of 70°F. If cold eggs are set without this warming period, water condenses on the shells, and yolks occasionally rupture.

For high-percentage hatches, it is *essential* that eggs are incubated in the correct position. Always set them on their sides with the large end (air cell) slightly raised. When duck eggs are set with the air cell lowered, their chances of hatching are decreased by up to 75 percent. Set only the number of eggs that fits comfortably in the tray, without crowding or stacking them on top of one another. If at all possible, do not disturb the eggs during the first 24 hours in the incubator.

Our Incubation Procedures

During the 40 years I have been raising ducks, the following procedures have been found to give excellent results. We have experimented with many alternatives, but these methods work best for us. The incubators we use are forced-air models, rated to hold approximately 2,500 chicken eggs in 12 trays. We use a 2,500-egg separate hatcher that is vented to the outdoors to reduce the amount of down and dust in the hatchery room. The incubator room is kept at 75 to 80°F during the hatching season.

1. Eggs are gathered in the morning and again in mid-afternoon. All duck eggs are washed the day they are laid in 100 to 110°F water to which an antibacterial soap has been added.

2. The eggs are stored in a basement egg room at a temperature of 55 to 65°F for 1 to 7 days.

3. Because eggs from some breeds (and even some varieties and strains within a variety) have somewhat different incubation periods, we set the eggs on a precise schedule in order to synchronize the hatch. Our goal is to have the ducklings ready to ship on Tuesday mornings. So, 4 weeks prior to the desired hatch date, we start setting eggs on Monday evening in the following order: Magpie, Swedish, Saxony, Runner (White, Black, Blue, Chocolate), Aylesbury, Rouen, Pekin, Cayuga, Appleyard, and Crested. First thing Tuesday morning we set Ancona, Harlequin, Muscovy, and other varieties of Runner eggs. Late Tuesday afternoon, the Campbell eggs are set. On Tuesday evening, the Bantam duck eggs are put in the hatchery room to start warming up, and then put in the incubator on Wednesday morning.

4. Only every other tray is filled the first week. Two weeks later, the other six trays are filled. This staggered setting prevents the incubator from overheating toward the end of the incubation period when the embryos are generating considerable heat.

5. From the 4th to 26th day of incubation, the eggs are lightly sprayed once daily with 100°F tap water.

6. From day 1 through day 25, the eggs are turned 90 degrees once an hour by an automatic turner.

7. From day 1 through day 25, the incubation dry-bulb temperature is kept at 99 to 99.25°F and the wet-bulb temperature is run at 83 to 84°F. The incubation season runs from February through June. Because eggshells gradually become more porous as the laying season progresses, we gradually increase the wet-bulb reading so that by the last setting in early June, the wet-bulb temperature is 86 to 87°F.

8. On Friday evening during the fourth week of incubation (the 24th day of incubation), the eggs are transferred from the incubator to the hatcher. The hatcher is operated at a dry-bulb temperature of 98.5°F and a wet-bulb temperature of 90°F. On Saturday evening, the eggs in the hatcher are lightly wetted with a fine spray of lukewarm tap water. Normally, this is the only time the hatcher door is opened after the eggs are transferred and prior to removing the ducklings.

9. On Sunday morning (the 26th day of incubation), the dry-bulb temperature is dropped to 98°F and the wet-bulb temperature raised to 92°F. As the hatch progresses, the wet-bulb temperature is kept at 92 to 94°F.

10. On Monday and Tuesday mornings, the hatcher is opened, the trays removed, and the ducklings taken out. The unhatched eggs are returned to the hatcher and lightly sprayed with lukewarm tap water to soften the shell membranes, which may have dried while out of the machine. On Wednesday morning, the remainder of the hatch is taken off. (If we were hatching ducklings not to be shipped, the hatcher would not be opened until Tuesday. Remember, every time the hatcher is opened the humidity level of the machine drops and shell membranes can dry out and become tough.)

11. On Wednesday afternoon, all removable components of the hatcher are taken outside and thoroughly washed with hot water and antibacterial soap. The hatcher cabinet is vacuumed and then thoroughly scrubbed with hot water and antibacterial soap. The hatchery room is vacuumed, all surfaces are wiped down with water and soap, and the hatcher put back together.

12. On Friday all vents of the hatcher are closed, the inside surfaces (including the hatching trays) are sprayed with quaternary ammonium disinfectant mixed with water, at three times the strength recommended for normal use. (I wear a respirator with charcoal filters during this procedure.) Once the inside is sprayed, the hatcher door is closed, the hatcher is turned on and allowed to run for 30 minutes with the vents closed. At the end of 30 minutes, the vents are opened wide and the machine allowed to dry out while the fan and heater are left on.

Temperature

In still-air machines the correct temperature is 101.5°F, 102°F, 102.5°F, and 103°F for each consecutive week. If you do not fill the incubator with one setting, but add a few eggs each week, a constant temperature of 102° to 102.5°F should be maintained. It is essential that thermometers be positioned properly in still-air incubators or an incorrect temperature reading will be given. The top of the thermometer's bulb must be level with the top of the eggs. *Do not* lay the thermometer on top of the eggs since this practice will give a warmer temperature reading than actually exists at the level of the eggs. Forced-air machines are maintained at several degrees lower, 99.25 to 99.75°F, since all sides of the egg are warmed equally.

The temperature must be watched closely at the last 7 to 10 days of incubation. During this period, an increase in temperature is often experienced, and the thermostat may need to be adjusted slightly each day to keep the eggs from overheating. Lowering the temperature by 1 to 1.5°F for the final 2 days is beneficial, since ducklings generate considerable internal heat in their struggle to free themselves from their shells.

It is a good idea to use thermometers designed specifically for incubators, as they have greater accuracy than utility models and are easier to read.

Humidity

To have a large number of strong ducklings hatch, the contents of the eggs must gradually dehydrate in the correct amount. When dehydration is excessive, the embryos are puny and weak, making it difficult for them to break out of the eggs. Conversely, inadequate moisture loss results in chubby embryos that have difficulty turning within the egg and cracking all the way around the shell.

The rate at which the contents of eggs dehydrate is regulated by the quantity of moisture in the air of the incubator and the porosity of the eggshell. Moisture in small incubators is usually supplied by water evaporation pans, whereas in large commercial machines, humidity is typically supplied by spray nozzles or evaporation panels.

The correct level of humidity varies widely depending on factors such as: elevation, breed of duck, egg size and shell quality, storage length and humidity prior to incubation, environmental temperature while egg is in the nest prior to being gathered, and length of time female has been laying eggs. (The longer a duck lays, the more porous her eggs become.) Therefore, it is impossible to give an exact recommendation. A good procedure in determining the correct humidity level is to start out by following the basic recommendations given in your incubator manual or those given here, and making necessary adjustments as needed.

Determining Correct Humidity Level

The following procedure is how we determine the correct level of humidity:

1. On a sheet of paper, we write down the incubator number and the date eggs were set, and prepare a column for wet-bulb readings and one for dry-bulb readings. Down the left side, we write the days of the month.

2. Every morning at approximately the same time, we read and record the dry-bulb and wet-bulb temperatures. (Be sure to take these readings before the machines are opened.)

3. At the end of the incubation period, each column is added up and divided by the number of days to get average readings.

4. If the hatch is satisfactory, the machine is operated at those average dry-bulb and wet-bulb readings for the next batch of eggs.

5. If the hatch was unsatisfactory, adjustments are made accordingly.

6. By keeping these records on hand, we have learned over time the intricacies of each machine for different seasons of the year and for different kinds of eggs.

If your incubator is equipped with a wet-bulb thermometer or hygrometer (if it isn't, it's difficult to know what the humidity is), the correct reading on these instruments during the incubation period usually ranges between 82 and 85°F on the wet-bulb thermometer, which is equal to a relative humidity of approximately 55 percent on the hygrometer. When eggs have been washed prior to incubation, the correct humidity level is normally between 85 and 88°F on the wet-bulb thermometer or 65 percent on the hygrometer. Throughout the hatch (the last 3 days of incubation), the relative humidity should be increased to approximately 75 percent, or 92 to 94°F on the wet-bulb thermometer.

The size of the air cell (normally located at the large end of the egg) is a useful indicator of whether the contents of the eggs are dehydrating at the correct rate. The air cell's volume can be observed by candling the eggs in a darkened room. On the 7th, 14th, 21st and 26th days of incubation, the average air cell volumes should be approximately the same size as those in the accompanying illustration (keeping in mind that the eggs of some breeds, strains, or individual females may show significant difference in air cell size and still hatch satisfactorily). If the air cells are too large, increase the moisture in the incubator and/or decrease the amount of ventilation, being careful not to reduce the airflow so severely as to suffocate the embryos. If the air cells are too small, decrease the moisture level and/or increase the ventilation.

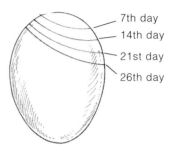

7th day
14th day
21st day
26th day

When candled, eggs that are dehydrating at the proper rate will show air cell sizes similar to those here.

Ventilation

The developing embryo needs a constant supply of fresh air, which is provided through vent openings in the sides and tops of incubators. The amount of ventilation required is relatively small, but essential to the well-being of the imprisoned ducklings. Ventilation demands increase at hatching time.

Turning during Incubation

Incubating eggs must be turned to exercise the embryo and relieve stresses. Some eggs will hatch if you turn them just once every 24 hours, but turning three times daily at approximately 8-hour intervals is the *minimum* for high-percentage hatches. For best results, duck eggs need to be rotated at regular hours and revolved at least one-third of the way around at each turning. Some studies indicate that turning eggs a full 180° improves hatchability. Eggs must be turned gently to avoid injury to the embryo. Turning should begin 24 to 36 hours after the eggs are set, and be discontinued 3 days before the scheduled hatch date.

When eggs are turned manually, it is helpful to mark them with an X and an O on opposite sides with a wax or lead pencil. Liquid inks — such as those in felt-tipped pens — should not be used, since they clog shell pores and can poison the embryo. After each turning, all eggs should have the same mark facing up.

Cooling

For best results when using still-air incubators, eggs should be cooled once daily — except during the 1st week and last 5 days of incubation. (Satisfactory hatches are obtained in forced-air machines without cooling.) When the room temperature is 65 to 70°F, the trays of eggs should be removed from the machine and cooled for 5 minutes a day the 2nd week, 8 minutes daily the 3rd week, and 12 minutes the first 4 days of the 4th week.

If you're like me, you may get sidetracked and forget the eggs as they cool. While they will hatch if left without heat for several hours once or twice during incubation (except during the 1st week and the last 5 days when low temperatures are disastrous), repeated over-cooling will retard growth and can be fatal. To prevent this, I set a timer for the appropriate number of minutes at the beginning of each cooling period.

If you ever find that the temperature in the incubator is excessively high by more than 2°F, *immediately* cool the eggs for 10 minutes and make adjustments to correct the problem.

Spraying or Sprinkling?

After considerable experimentation, we have found that we get consistently higher-percentage hatches by lightly spraying duck eggs once a day with lukewarm tap water from the 4th to the 26th day of incubation. Eggs that are sprayed regularly dehydrate more than unsprayed eggs.

To prevent the egg membranes from drying out and becoming tough during the hatch, it is sometimes necessary to spray or sprinkle duck eggs with warm water 48 and again 24 hours before the calculated hatching time.

Candling

The best time to candle duck eggs to check fertility is on the 5th to 7th days of incubation. If eggs are candled prematurely, it is more likely that fertile eggs will be missed and accidentally discarded.

Eggs are candled in a darkened room with an egg candler or flashlight. On the 7th day, fertile eggs reveal a small dark spot with a network of blood vessels branching out from it, closely resembling a spider in the center of its web. Infertile eggs are clear with the yolk appearing as a floating shadow when the egg is moved from side to side.

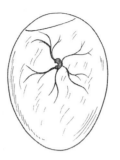

On the 7th day, a fertile egg (left) has a spidery network of blood vessels, while an infertile egg (right) is clear.

Sometimes embryos begin to develop but perish within several days. When this happens, a streak or circle of blood is visible in an otherwise clear egg. Contaminated and rotten eggs often exhibit black spots on the inside of the shell, with darkened, cloudy areas floating in the egg's interior. All eggs not containing live embryos should be removed from the incubator.

Contaminated and rotting eggs give off harmful gases and frequently explode, covering the other eggs and the incubator's interior with putrid-smelling, bacteria-laden goo that is difficult to clean up. To reduce the chances of blow-outs, duck eggs should be candled on the 12th, 19th, and 24th days of incubation.

A blood ring visible when candling indicates that the embryo has died.

The Hatch

All the time and effort invested in producing and incubating eggs is rewarded by the hatch. Those first muffled chirps of the ducklings are sweet music.

Normally the first eggs will be pipped, or broken through, 48 hours prior to the hatch date. Ducklings require 24 to 48 hours to completely rim the shell and exit. Newly hatched birds are wet and exhausted, and should remain in the incubator 4 to 12 hours while gaining strength and drying off.

If your incubator is equipped with adjustable air vents, they should be regulated to give hatching ducklings additional air. However, do not open them so wide that an excessive amount of warm air escapes and the humidity drops. During the hatch, you may find it necessary to place extra water containers in the small machines to maintain an adequate level of humidity with the increased air circulation. Evaporation pans should be covered with screen or hardware cloth to make *certain* that the ducklings cannot drown.

After most of the ducklings are hatched, the relative humidity can be lowered to 50 percent (82 to 84°F on the wet-bulb thermometer) so they will fluff out properly.

Removing Ducklings from the Incubator

When the ducklings are dry, remove them from the incubator. Before opening the machine, prepare a clean container — with sides at least 6 inches high — with soft bedding. While transferring ducklings, work quickly and gently, discarding empty shells and pipped eggs containing birds that are obviously dead. The room temperature should not be below 70°F. Any wet ducklings can be left in the incubator for several more hours.

Separate Hatcher Highly Recommended

Rather than incubating and hatching eggs in the same machine, there are advantages to incubating eggs in an incubator for the first 23 to 24 days and then transferring them to a separate machine for the last few days during the hatch. By using a separate hatcher, the unique turning, temperature, and humidity requirements of eggs at different stages can be better met, the incubator is kept cleaner, and the hatcher can be thoroughly cleaned and disinfected after each hatch.

Help-Outs

At the end of the hatch some live ducklings may still be imprisoned within partially opened shells. Assistance can be given by carefully breaking away the shell just enough so that the duckling will be able to exit.

While some birds that are assisted from the shell develop into fine specimens, a large percentage of them are handicapped by a deformity or weakness. When it is understood that the hatch is a fitness test given by nature to cull out the weak and deformed, we can take a more realistic view of helping ducklings from the shell. It is a good idea to mark all help-outs so they are not used as breeders unless they exhibit exceptional qualities as mature birds.

Incubation Checklist

❑ Provide adequate nests furnished with clean nesting material.

❑ Gather eggs in morning to protect them from temperature extremes.

❑ Handle eggs with clean hands or gloves and avoid shaking, jolting, or rolling them.

❑ Store eggs in clean containers in a cool, humid location, and turn them daily if held longer than 5 days.

❑ Set only clean (or reasonably clean) eggs with strong shells.

❑ Start incubator well in advance of using. Make adjustments of temperature, humidity, and ventilation before eggs are set.

❑ When filling the incubator, position eggs with their air cells slightly raised, and do not crowd eggs on tray.

❑ Try not to disturb eggs during the first 24 hours in the incubator.

❑ Gently turn eggs at least three times daily at regular intervals from the 2nd to 25th day (2nd to 32nd for Muscovies).

❑ Operate still-air incubators at 101.5 to 103°F, and forced-air machines at 99.25 to 99.75°F during the incubation period.

❑ Maintain relative humidity at 55 percent (a wet-bulb reading of 84 to 86°F) for first 25 days (32 days for Muscovies) of incubation.

❑ When using a still-air machine, cool eggs 5 minutes a day the 2nd week, 8 minutes daily the 3rd week, and 12 minutes the first 4 days of the 4th week.

❑ Lower the incubator temperature 1 to 1.5°F for the hatch.

❑ Increase the relative humidity to 75 percent (a wet-bulb reading of 90 to 94°F) for the hatch.

❑ Sprinkle eggs with lukewarm water 48 and 24 hours prior to the scheduled hatch date.

❑ Do not open incubator except when necessary during the hatch.

❑ Leave ducklings in machine until they are dried.

❑ Clean and disinfect incubator after each hatch.

Incubator Sanitation

While the incubator supplies the correct conditions for the embryo to develop, it also provides an excellent environment for the rapid growth of molds and bacteria. Consequently, it is essential that the incubator be kept as clean as possible.

At the conclusion of *each* hatch, clean and disinfect the incubator. If the eggs are set to hatch at various times, the water pans should be emptied, disinfected, and returned with clean, warm water, and the duckling fuzz removed from the incubator with a damp cloth or vacuum cleaner.

At the end of the hatching season, thoroughly clean the incubator and store it in a dry, sanitary location.

Buried Treasures

For my 10th birthday, Mom asked if there was something special I'd like to do. Without hesitation, I told her I wanted to go to the city park at Waverly Lake to feed the ducks. Mom knew that there was no use in pointing out that we had a pasture full of ducks that I fed every day.

On a beautiful Sunday afternoon my family joined in feeding stale bread and grain to my eager web-footed friends. When the larder ran out, we explored the shoreline. The previous week had brought torrential rains, and the lake was the fullest in memory. As we walked the shore, I spotted a nest of eggs lying a foot below the murky water's surface. They must be rescued! Dad voted for leaving them, pointing out they had probably been submerged for nearly a week, but the 13 pale green eggs were soon nestled in an old towel.

At home I placed the precious eggs in my incubator. The days passed slowly. On the 27th day, I heard peeping. Lifting the lid, I saw eight black-and-yellow "chipmunk"-marked Mallards. The smallest one was christened Tiny Tina and quickly became my favorite duck.

Pinpointing Incubation Problems

SYMPTOMS	COMMON CAUSES	REMEDIES
More than 10 or 15% clear eggs when candled on 7th day of incubation	Too few or too many drakes	Correct drake-to-hen ratio
	Old, crippled, or fat breeders	Young, active, semi-fat breeders
	Immature breeders	Use breeders 7 months or older
	Breeders frequently disturbed	Work calmly around breeders
	No swimming water for mating	Swimming water for large breeds
	First eggs of the season	Don't set eggs laid 1st week
	Late-season eggs	Don't set eggs when males molt
	Medication in feed or water	Avoid medicating breeders
	Eggs stored more than 14 days	Set fresher eggs
Blood rings on 7th to 10th day	Faulty storage of eggs	Proper storage prior to setting
	Irregular incubation temperature	Adjust machine ahead of time
Ruptured air cell	Rough handling or a deformity	Handle and turn eggs gently
Yolk stuck to shell interior	Old eggs that haven't been turned regularly during storage	Turn eggs daily if they are held for more than 5 days
Dark blotches on shell interior	Dirt or bacteria on shells causing contamination of the inner egg	Wash eggs with water and disinfectant soon after gathering
More than 5% dead embryos between 7th and 25th day of incubation	Inadequate breeder diet	Supply balanced diet
	Highly inbred breeding stock	Introduce new birds to flock
	Incorrect incubation temperature	Check accuracy and position of thermometer
	Periods of low or high temperature	Check temperature often; don't overcool eggs
	Faulty turning during incubation	Turn at least three times daily

SYMPTOMS	COMMON CAUSES	REMEDIES
Early hatches	High incubation temperature	Lower incubation temperature
Late hatches	Low incubation temperature	Raise incubation temperature
Eggs pip but do not hatch; many fully developed ducklings dead in shells that aren't pipped	High humidity during incubation	Decrease amount of moisture
	Low humidity during incubation	Increase amount of moisture
	Eggs chilled or overheated during last 5 days of incubation	Protect eggs from temperature extremes during this period
	Low humidity during the hatch causing egg membrane to dry out	Last 3 days, raise humidity to 75% and sprinkle eggs daily
	Poor ventilation during hatch	Increase air-flow during hatch
	Disturbances during hatch	Leave incubator closed
Eggs pipped in small end	Eggs incubated in wrong position	Position eggs with small end lower than large end
Sticky ducklings	Probably low humidity during incubation and/or hatch	Increase amount of moisture
Large, protruding navels	High temperature	Lower temperature; check thermometer
	Excessive dehydration of eggs	Increase humidity level
	Bacterial infection	Improve sanitation practices
Dead ducklings in incubator	Suffocation or overheating	Increase ventilation during the hatch and watch temperature carefully
Spraddled legs	Smooth incubator trays	Cover trays with hardware cloth
More than 5% cripples other than spraddled legs	Inadequate turning during incubation	Turn eggs a minimum of 3 times daily
	Prolonged periods of cooling	Don't forget eggs when cooling
	Inherited defects	Select breeders free of defects

THIRTEEN

REARING DUCKLINGS

The downy young of all types of poultry are charming, and day-old ducklings are no exception with their bright eyes, tiny wings, and miniature webbed feet. Aided by shovel-like bills, they are soon eating an amazing quantity of food, all the while growing at an equally astounding rate. In only 8 to 12 weeks, a newly hatched duckling is transformed into a young adult.

Basic Guidelines

Of all domestic fowl, the young of ducks are the easiest to raise. It is common for every duckling in a brood to be reared to adulthood without a single mortality. They can withstand less than optimum conditions well — although that is no reason to mistreat or neglect them. If you put into practice the following management guidelines, raising ducklings will be a pleasant and trouble-free task.

   1. **Keep them warm and dry and protect them from drafts.** Ducklings that are cold lose their appetites, and when wet, they chill rapidly and will die if not dried and warmed promptly. Drinking-water containers should be designed so that ducklings cannot enter them and become excessively wet.

   2. **Maintain them on dry bedding that provides good footing.** Many internal parasites, molds, and disease organisms thrive in damp and filthy bedding. Smooth, slick floors are the leading cause of spraddled legs.

   3. **Supply fresh, nonmedicated feed that provides a balanced diet.** A proper diet that is formulated to meet all of the dietary requirements of ducklings is essential for disease resistance and normal growth. Feed that is moldy, deficient in nutrients, or contains certain additives can cause

stunted growth, sickness, or death. Never use feed containing arsenicals or the drug roxarsone. (If nonmedicated feeds are unavailable in your locale, rations containing amprolium, chlortetracycline, neomycin, oxytetracycline, sulfaquinoxaline, or zinc bacitracin have been used successfully by many duck growers and should work satisfactorily for you if one of these drugs is incorporated in the feed at the proper dosage. See Poisoning from Medication, chapter 17, page 265.)

4. Provide a constant supply of fresh water. Ducklings suffer a great deal when drinking water is not available, particularly right after they have eaten. (Never add Ren-O-Sal tablets to the drinking water of ducklings — the active ingredient roxarsone can be deadly to young waterfowl.)

5. Furnish adequate floor space and fresh air. Forcing ducklings to live in crowded conditions is one of the leading causes of feather eating, wing disorders, and disease.

6. Protect them from predators. Tame and wild predatory animals and birds find unprotected ducklings easy prey. The clumsy feet of humans and large animals also snuff out the lives of many young birds.

Natural Brooding

For the home poultry flock, natural brooding can have many advantages. It eliminates purchasing or constructing special brooding equipment and supplying artificial heat. When allowed to range, hen-mothered ducklings learn to forage for their food at a young age, and will be nearly self-sufficient if there is a plentiful supply of insects, tender grass, and wild seeds and fruits. Ducklings can be brooded by duck, turkey, bantam, or large chicken hens.

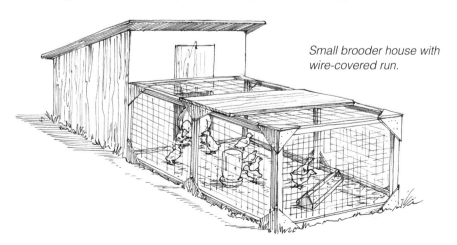

Small brooder house with wire-covered run.

A day-old Mallard exhibiting the bold black and yellow markings that are typical of all gray varieties of ducklings.

By 2 weeks of age, the down color has faded and feathers can be felt protruding from her tail and sides.

At 4 weeks, feathers cover the face, shoulders, underbody, and tail and have begun to appear on her back.

Now 8 weeks old, the once soft, cuddly duckling is fully feathered and nearly the size of her parents.

Managing the Hen and Her Brood

The best management practice in dealing with a hen and her brood of ducklings is to bother them as *little* as possible. Your main concern is to protect them from predators and to keep the ducklings from becoming soaked during their first several weeks of life outside the shell.

Free-roaming turkeys and chickens often leave ducklings stranded, so it is a good idea to enclose foster hens with their broods in a dry building or pen for the first 2 to 4 weeks to protect them from rain, cold winds, predators, and rodents. Concentrated feed can be supplied to get them off to a fast start.

The same procedures are recommended for duck hens and their broods, although it is normally safe to give them freedom in a large pen not occupied by other fowl after the ducklings are several days old. Duck hens are not as likely to jump or fly over barriers such as fences and leave the little ones behind.

Confine hens and their broods in a secure pen or building each night until the ducklings are 6 to 8 weeks old. This significantly reduces the possibility of losses to predators.

Hens can be aggressive toward strange ducklings and may injure or kill them. Don't confine two or more hens together with their broods, particularly during the first week or two.

Allowing a new brood in with an established flock can also be dangerous for the young ones, especially

Most children love to hold baby ducks. To prevent injury from dropping or suffocation, ducklings should be held on the palm of a cupped hand while the second hand is placed over the top, as demonstrated here by our 7-year-old nephew, Christopher.

when ducks are penned in close confinement and the strange ducklings may threaten the territorial boundaries of the adults. If your flock of ducks is permitted to roam freely, ducklings may not be bothered. In any case, newly hatched ducklings should be allowed with adult birds *only* if you can be on hand to remove them if a problem develops.

Giving Hens Foster Ducklings

Sometimes it is desirable to give hens foster ducklings to brood. If a hen already has young, the foster ducklings should be approximately the same size and color as her own. Hens who set but for one reason or another do not hatch any ducklings — or chicks in the case of a chicken hen — often will accept a foster brood. Giving ducklings to a chicken hen that has chicks of her own usually does not work out, but it can be attempted in an emergency situation. To lessen the possibility of rejection, it is advisable to slip foster ducklings under their new mother after dark.

Novice duck raisers sometimes lock a hen (that has been setting for a short time or not at all) in a pen with ducklings and expect her to mother them. Almost without exception these attempts fail and frequently result in ducklings being brutally attacked. Unless an exceptional hen is discovered, it is not wise to give foster ducklings to hens that have not set their full term.

The number of ducklings a hen can brood depends on her size and the weather conditions. Typically, a duck hen can successfully brood 8 to 12 of her own young. Bantam hens can handle 6 to 8 ducklings, large chickens 12 to 18, and turkey hens up to 25. The most important factor is that all of the ducklings can be hovered and warmed at the same time.

Artificial Brooding

If a broody hen is not available or you are raising a large number of birds, it will be necessary to brood ducklings artificially. When sound management practices are used, this method produces good results.

The Brooder

Brooders provide warmth for the ducklings during the first 4 to 6 weeks after they hatch. Brooding equipment can be purchased from stores handling poultry supplies or can be fabricated at home with a minimum of skill and cost.

Battery Brooder

This brooder is an all-metal cage equipped with a wire floor, removable dropping pan, thermostat, electric heating element, and feed and water troughs. They can be purchased and used in individual units or stacked on top of one another. Battery brooders are commonly used for starting chicks, but can also be used for brooding ducklings up to an age of 2 to 6 weeks, depending on the breed. While new units are expensive, this type of brooder requires limited floor space, is easily cleaned, protects young birds from predators, and provides sanitary conditions.

Hover Brooder

Available with gas, oil, or electric heat, this brooder consists of a thermostatically controlled heater that is covered with a canopy. The brooder is supported with adjustable legs or suspended from the ceiling with a rope or chain and is easily set to the proper height as the birds grow. Water fountains

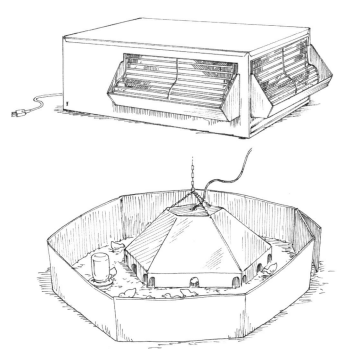

A battery brooder pro-
vides heat, troughs for
food and water, and good
protection from predators.

When hung from a chain,
a hover brooder — shown
here with draft guard —
can be raised as duck-
lings grow in size.

and feeders are placed around the outside of the canopy where the ducklings
drink, eat, and exercise in cooler temperatures. Hover brooders are available
in sizes that are rated at 100- to 1,000-chick capacity. The number of ducklings under each unit should be limited to 50 to 60 percent of the given
capacity for chicks.

Heat Lamp

A simple method for brooding ducklings is with 250-watt heat lamps.
Suspended 18 to 24 inches above the litter, each lamp will provide adequate
heat for 20 to 40 ducklings, depending upon the size of birds and the outside
temperature. We always use at least two lamps just in case one burns out.

When heat lamps are used, *extreme care* must be taken to prevent fires.
These lamps must *never* be hung with the bottom of the bulb closer than 18
inches from the litter, or so that any part of the bulb is near flammable material such as wood or cardboard. *Always* make sure heat lamps or other types of
heaters are securely fastened to something that is sturdy and will not fall over
or collapse. *Always* use porcelain light fixtures with heat lamps — never plastic. *Always* make sure heat lamps are positioned so that the bulb is a minimum
of 18 inches away from the bedding or sides of a combustible brooding frame.

Homemade Brooder

A few ducklings can be brooded with a light bulb that is positioned in a box or wire cage. Research indicates that blue bulbs are best since they reduce the incidence of feather eating and are gentler on the ducklings' eyes.

The wattage required depends on the size of the box and the room temperature. I prefer to use several 40-watt bulbs rather than a single larger one. The box or cage must be big enough to allow ducklings to move away from the heat when they desire. When bulbs of over 40 watts are used, they must be located out of the reach of ducklings. To prevent fires or the asphyxiation of ducklings from some smoke produced by smoldering materials, even bulbs of low wattage *must not touch* or be close to flammable substances.

Our Brooding and Rearing Procedures

There are a variety of ways that ducklings can be brooded and reared successfully. The procedure used will depend on factors such as the number of ducklings to be raised, the facilities available, climate, the purpose of the ducks (pets, show, breeding stock, laying stock, broilers for meat, or roasting birds), and personal preferences.

Our main purpose in raising ducks is to help preserve and distribute heritage breeds, varieties, and strains. Therefore, we have devised a system that produces ducks that are healthy, physically vigorous, productive, long-lived, and safe from predators. We also try hard to provide an environment in which our ducks can enjoy themselves. As early as it becomes safe, they spend their days outside in grassy yards or pastures where they can enjoy natural foods, fresh air, sunshine, and plenty of exercise.

1. Approximately 12 hours after hatching, the ducklings are removed from the hatcher, placed in chick shipping boxes, and left in the 70 to 80°F hatchery room for 36 to 48 hours.

2. Two days after hatching, the ducklings are put in a battery-type turkey brooder (with wire floor) that has been preheated to 95°F. The bill of each duckling is dipped into the lukewarm drinking water (to which ⅛ level teaspoon of V.M.P. Poultry Pac by R.X.V. Products per gallon of water has been added — the V.M.P. stands for vitamins, minerals, and probiotics). In addition to the water trough along the front of the brooder, we place a quart waterer inside for each 15 to 20 ducklings to ensure that they will readily

A light reflector and clamp make a safe and inexpensive hover brooder. They are available at hardware stores and, when not used for brooding, come in handy around the shop and house. The reflector can be clamped to the side of a cage or box and will brood up to a dozen ducklings. As the young ones grow, the heat source can be adjusted to the correct height.

A variety of homemade brooders can also be crafted. Hover-type brooders can be built with a plywood or sheet-metal canopy and porcelain light fixtures. Better still is the washtub brooder built and used by a retired coal-miner friend of ours. He outfitted an old zinc washtub with several light fixtures and three adjustable legs. With just a little work and a lot of ingenuity, he created a safe brooder that gave dependable service for many years.

find the water during the first 2 days. I also cut up little bits of grass and sprinkle it on the water to further entice them to drink.

3. The trough feeder is filled to the top with 18-percent-protein waterfowl starter/grower crumbles. Chick-sized granite grit and old-fashioned oatmeal (uncooked) are lightly sprinkled on top of the crumbles.

4. At 5 to 10 days of age (depending on the weather) the ducklings are placed in a brooder building having wooden floors covered with 1½ inches of coarse cedar sawdust. Heat is provided by 250-watt heat lamps 18 inches above the floor. Feed is supplied in hanging feeders and water in 5-gallon vacuum waterers located on a outside water porch (that has a floor covered with welded wire with openings ½ inch x 1 inch in size). For the first several days after being moved to the brooder building, we continue using the V.M.P. Poultry Pac in the drinking water.

5. Chopped lettuce, dandelion greens, or tender young grass is fed every day.

6. Starting at approximately 2 weeks of age, 16-percent-protein grower pellets (that include 20 percent oats) are gradually introduced by mixing with the starter crumbles over the course of several days.

7. As soon as the weather permits, the ducklings are allowed access to grassy yards. The heat lamps are left on in the brooder building for the ducklings to use when they get cool. Adult geese are kept around the perimeter of these yards to discourage airborne or ground predators.

8. Every night and during daytime inclement weather, the ducklings are walked into the brooder building and locked in.

Draft Guard

The use of a draft guard around hover brooders and heat lamps is recommended for the first week or two. The guard protects ducklings from harmful drafts and prevents them from wandering too far from the heat and piling up in corners. Commercially produced guards constructed of corrugated cardboard can be purchased and used several times. Homemade guards can be fashioned with 12-inch boards or welded wire that is covered with burlap or paper feed sacks. The guard should form a circle 2 or 3 feet from the outside edge of the hover brooder canopy.

Heat

Ducklings require less heat than chicks. Under the brooder the temperature should be held at approximately 90°F the first 7 days, and then lowered 5°F each successive week. Once ducklings are 6 to 8 weeks old and well feathered, they can withstand temperatures down to 50°F or lower, but must be protected from drastic temperature fluctuations.

The actions of the ducklings are a better guide to the correct temperature than a thermometer. If ducklings are noisy and huddle together under the heat source, they are cold and additional heat should be supplied. When they stay away from the heat, or pant, they are too warm and the temperature needs to be lowered. The proper amount of heat is being provided when ducklings sleep peacefully under the brooder or move about freely, eating and drinking.

Even at the start of the brooding period, it is *extremely* important that ducklings are able to get away from the heat source when they desire. Overheating is almost as damaging to ducklings as chilling.

Brooding without Artificial Heat

If you don't have electricity available for artificial heat, you can still brood ducklings. One technique is to keep ducklings in a box near the stove or furnace until they are large enough to be put outside. If you use this method, be careful not to place the container too close to the heat, which could overheat the ducklings or, worse yet, start a fire.

Another procedure that works satisfactorily in mild weather is to utilize the body heat given off by the ducklings to warm themselves. A well-insulated box is the basic equipment needed. The floor of the container should be covered with 2 to 4 inches of dry bedding such as clean rags, chopped grass, straw, or sawdust. If a fine bedding such as sawdust is used, cover it with cloth or burlap the first several days to prevent ducklings from harming themselves by eating the particles of wood.

The top and sides of the box must be draped with a layer of old towels, blankets, or burlap bags. Be sure that the little ones have sufficient air to prevent suffocation. Since heat rises, the ducklings will stay warmer if the "hot box" you design has a maximum of 6 inches of headroom after the bedding is in place. Also keep in mind that the smaller the inside area is, the cozier the occupants will be.

A small door should be made in the side of the brooder box where the ducklings can exit to eat and drink. They soon learn to go back inside when they're chilly, although you'll probably need to give them a helping hand the first day or two.

Floor Space

When ducklings of medium- to large-sized breeds are started on litter, allow a minimum of .75 square feet of floor space per bird for the first 2 weeks, 1.75 square feet until 4 weeks of age, 2.75 square feet until 6 weeks, and 3 to 5 square feet per bird thereafter. At 3 to 4 weeks of age, give ducklings access to an outside pen or yard during mild weather, allowing 10 square feet of space per bird. This additional space will help keep the inside bedding drier and reduce sanitation problems. Bantam breeds require approximately one-half the floor space of large breeds.

Litter

Ducklings consume a huge volume of water so it is *exceedingly important* that a thick layer of absorbent, mold-free litter (or wire flooring) is used to keep the brooding area from becoming sloppy. Shavings, sawdust, peanut hulls, peat moss, crushed corncobs, flax, or chipped straw can be used for bedding.

Begin with 3 to 6 inches and add new litter as required. Soggy and caked bedding should be removed whenever it appears and replaced with dry material. A daily stirring of litter is advantageous, particularly if the density of ducklings is high. In warm weather, flies will become a problem if the quality of the litter is allowed to deteriorate.

Yards and Pastures

Ducklings can be given access to a yard or pasture as soon as they are comfortable outside. The exercise, fresh air, and sunlight are beneficial to their health. They will also eat substantial quantities of tender grass, which reduces their feed consumption and enriches their diets with vitamins and minerals. Ducklings *cannot* eat mature, dry, or coarse grasses.

When allowed outside, ducklings should be put under cover each time it rains until they are 5 to 6 weeks old. After this age, a simple shelter is adequate protection.

Water

Being waterfowl, ducklings love to swim and they consume three to four times more water than chickens do. In an effort to keep water receptacles clean, people sometimes add common household bleach to the drinking and swimming water of their ducks. This can be a dangerous practice because bleach can ravage the desirable bacteria in the birds' digestive tracts.

Swimming Water

It is not necessary to have swimming water for ducklings, even though they thoroughly enjoy going for a paddle within days after hatching. Even when ducklings are brooded by a duck hen, it is safest to keep them out of water until they are at least 2 weeks old.

To protect ducklings from drowning, all water containers that they can enter should have gently sloping sides with good footing to allow tired and wet swimmers to exit easily. If you supply swimming water in receptacles having steep or slick sides, an exit ramp *must* be provided if drowning losses are to be avoided. Drowning and becoming soaked and chilled while swimming are the leading causes of duckling mortalities in home duck flocks.

Drinking Water

To thrive, ducklings must have a *constant* supply of drinking water. Drinking fountains should be designed so that young ducklings *cannot* get into the water. To lessen the possibility of ducklings choking to death, the

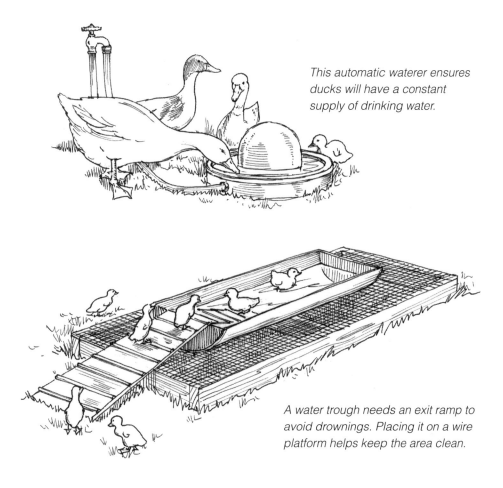

This automatic waterer ensures ducks will have a constant supply of drinking water.

A water trough needs an exit ramp to avoid drownings. Placing it on a wire platform helps keep the area clean.

water should be deep enough to allow them to submerge their bills and dislodge particles of food that frequently become stuck in their throats and nostrils.

Sufficient receptacles should be provided so their contents are not quickly exhausted and the ducklings left without water. Placing water containers on screen-covered platforms is a big help in keeping the watering area dry and sanitary. Waterers should be rinsed out daily.

If maximum growth rate is not required, once ducklings are several weeks old, they can go for 8 to 10 hours at night without water *if* they do not have access to feed.

Nutrition

The importance of a sound feeding program cannot be overemphasized. Nutrition is probably the most neglected phase of management in many small home flocks. Nearly half of the small duck flocks I observe exhibit nutritional deficiencies that can be avoided if the following recommendations are implemented.

The rate at which ducklings grow is in direct proportion to the *quantity* and *quality* of the feed they consume. For maximum growth they need a diet that provides 20 to 22 percent protein up to 2 weeks of age, and 16 to 18 percent protein from 2 to 12 weeks.

To stimulate fast growth, ducklings should be allowed to eat all the feed they want up to 2 weeks of age. After this time, you can limit them to 2 or 3 feedings daily, when they should be given all the feed they can clean up in 5 to 10 minutes. When month-old ducklings have access to succulent pasture, they can be limited to one feeding daily. Giving the birds their meal in the evenings will encourage them to forage throughout the day.

Suggested Feeding Schedule for Ducklings

TYPE OF DUCKLING	0–2 WEEKS LBS OF 18–20% STARTER FEED PER BIRD DAILY	2–7 WEEKS LBS OF 16–18% GROWER FEED PER BIRD DAILY	7–20 WEEKS LBS OF 15–16% DEVELOPER FEED PER BIRD DAILY
"Green"	Free choice	Free choice	—
Small Breed	Free choice	Free choice 5 min. twice daily	0.15–0.25
Egg Breed	Free choice	Same as above	0.20–0.30
Medium Breed	Free choice	Same as above	0.25–0.35
Large Breed	Free choice	Same as above	0.30–0.40
Muscovy	Free choice	Same as above	0.20–0.40

Note: The quantity of feed required by ducklings is highly dependent on the availability of natural foods, climatic conditions, and the quality of feed (e.g., birds require larger amounts of high-fiber foods than low-fiber foods to meet their energy requirements).

For the first several weeks, small-pelleted (³⁄₃₂ inch) or coarse crumbled feed is preferred; thereafter, larger pellets (³⁄₁₆ inch) will give the best results. Ducklings choke on fine, powdery mash when fed dry and up to 25 percent of the feed is wasted. Finely ground feed is better utilized when it is moistened with water or milk to a consistency that will form a crumbly ball when compressed in your hand. A new batch should be mixed up at each feeding to avoid spoilage and food poisoning.

Feeding Programs

The feed program you employ should be designed to fit your situation and goals, and most likely will be a combination of two or more of the following options.

Natural

Ducklings, particularly Muscovies and Mallards, which are brooded by their natural mothers, are capable of foraging for most of their own ration if there is an abundant supply of insects, wild seeds, and succulent plants. Ponds, lakes, sloughs, marshes, and slow-moving brooks are excellent sources of free food for ducklings. In most situations it is advisable to supply ducklings with concentrated feed for at least the first 10 to 14 days to get them off to a good start.

Grains

Small whole grains such as wheat, milo, kafir, or cracked corn can be fed to ducklings after they are several weeks old. Grains by themselves are *not* a balanced diet. Ducklings need tender greens and an abundant supply of insects or another protein supplement. They cannot be expected to remain healthy and grow well on a diet consisting exclusively of whole, cracked, or rolled grains.

Home-Mixed

It may be practical to mix your ducks' feed if the various ingredients are available at a reasonable price. The formulas given in the tables in appendix A on pages 284–285, Home-Mixed Starting Rations and Home-Mixed

Growing Rations, are examples of the types of feed that can be mixed at home. While these rations are not as sophisticated as commercially prepared feeds, they will give good results in most situations *if* they are mixed properly and the instructions are closely followed. If a vitamin-mineral premix (carried by many feed mills) is used, the rations provided in the table on page 286, Complete Rations for Ducklings (Pelleted), can also be home-blended.

Special equipment is not needed to mix duck feed. For small quantities the ingredients can be placed in a large tub and combined with the hands or a stick. Another method is to pile the measured components in layers on a clean floor and mix with a shovel. A cement mixer or old barrel mounted on a stand and outfitted with a handle, door, and ball bearings can be used for larger quantities. More important than the method used for mixing is that the ingredients are blended *thoroughly*. During warm weather, no more than a 4-week (3 weeks or less if ground grains are used) supply of feed should be prepared at a time.

Commercial

In some localities premixed starter and grower feeds for ducklings are available. When these feeds are used, the instructions should be followed. If rations specifically formulated for ducklings are not available, use the corresponding nonmedicated mixes recommended for game birds, turkeys, or chickens. The first two are preferred, but chicken feeds usually give satisfactory results if they are fortified with additional niacin.

You may want to have a local feed mill mix your feed if enough ducklings are raised to make it practical. The formulas given in the table on page 286, Complete Rations for Ducklings (Pelleted), are for complete rations that will provide a balanced diet *if* mixed properly and not stored for more than 4 weeks (less in hot weather). Ration numbers 9 and 10 are to be fed to ducklings up to 2 weeks of age, and numbers 11 and 12 from 2 to 12 weeks.

Niacin Requirements

Young waterfowl require two or three times more niacin in their diet than chicks. (See Niacin Deficiency, page 258.) When ducklings are raised in confinement on a ration that is deficient in niacin — such as commercial chick feeds — a niacin supplement should be added to their feed or water.

Niacin can be purchased in tablet form at drugstores, and is a common ingredient in poultry vitamin mixes. Adding 5 to 7.5 pounds of livestock grade brewer's yeast per 100 pounds of chicken feed (or 2 to 3 cups of yeast per 10 pounds of feed) will also prevent a niacin deficiency in ducklings.

Green Feed

The daily feeding of leafy greens to ducklings fortifies their diets with essential vitamins and minerals, reduces feed consumption, and lowers the possibility of cannibalism. Tender young grass (before it joints), lettuce, chard, endive, watercress, and dandelion leaves are excellent green feeds.

Ducklings will eat their salad *only* if it is tender and fresh. When greens are placed on the floor of a brooder, they soon wilt and are trampled and soiled. By putting the chopped feed in the ducklings' water trough, it remains succulent and clean, and the ducklings spend many contented hours dabbling for the bits of greenery.

Grit

Coarse sand or chick-sized granite grit should be kept before ducklings at all times. Grit aids the gizzard in grinding, helping birds to get the most out of their feed.

Feeders

For the first day or two, feed should be placed in containers where the ducklings cannot help but find it. Jar lids, shallow cans, and egg flats are excellent for this purpose. Once the ducklings have located their feed and are eating well, they should be fed in trough feeders to reduce wastage.

Feeders can be purchased or constructed at home out of scrap materials. Homemade troughs should be designed so they do not tip over as ducklings jockey for eating space. To keep birds out of the feeder, a spinner can be attached across the top.

Sufficient feeder space needs to be provided to ensure that each duckling receives its share. When limiting feedings to one or two daily, supply adequate trough space so that each bird can eat without having to struggle for its portion of the meal. By filling troughs no more than *half full*, you can significantly reduce the amount of feed that is wasted.

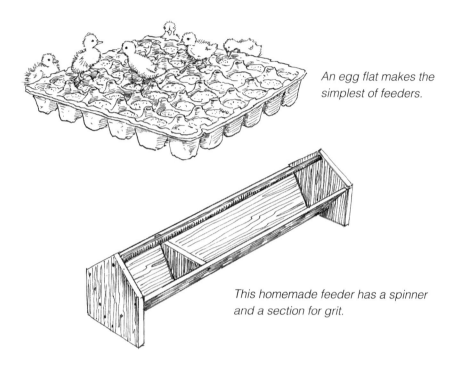

An egg flat makes the simplest of feeders.

This homemade feeder has a spinner and a section for grit.

Creep Feeding

When ducklings are brooded by hens and run with the flock, it is sometimes desirable to feed the young birds separately from the adult ducks. This is done by devising a creep feeder, where ducklings can eat without competing with grown birds.

The basic component of the creep feeder is a doorway large enough for only ducklings to pass through and gain access to the feed trough. The doorway panel can be placed across a corner of the duck yard or in the entrance of a shed.

Mature ducks can squeeze through smaller holes than is generally realized, and I have found that there is a tendency to make the slots too large. Dimensions for the portals will vary according to the breed raised, and you will probably need to do a little experimenting to find what size works best for your ducks. However, as a general rule, a space 4 inches square is satisfactory when breeds weighing 7 to 9 pounds are kept. For smaller breeds, such as Mallards, the passageway needs to be approximately 3 inches square to keep the old birds from entering and chowing with the youngsters.

Managing "Green" Ducklings

"Green" ducklings are the equivalent of broilers in chickens. These birds are managed to produce the quickest possible growth in the shortest period of time on the least quantity of feed. The breed best suited for this practice is the Pekin. With good management, they are capable of weighing 8 pounds at 7 weeks of age on approximately 20 pounds of feed.

To stimulate fast growth, "green" ducklings must have limited exercise, a continuous supply of high-energy, concentrated feed, and 24 hours of light daily. They are ready to be butchered as soon as their primary wing feathers are developed. If held beyond 7 or 8 weeks, their feed conversion decreases *rapidly* (see table on page 184) and they'll commence the molt, making it hard to pick them until they are 14 to 16 weeks of age.

Producing Lean Ducklings for Meat

The major drawback of raising ducks for meat is the high fat content of their carcasses. While many people enjoy the succulency of quick-grown duckling meat, research strongly indicates that many of us need to reduce the amount of fat in our diets. There are several management practices that can be used to produce leaner ducklings.

One method is to use a high-protein grower ration that contains an energy:protein ratio of 65:1 or less. While ducklings fed such rations do not gain *weight* as rapidly as birds given feed with a wider energy:protein ratio (88:1 is normally recommended), the actual amount of *meat* will be almost identical.

A second method, which is more effective and practical for the small-flock owner, is to raise a breed other than Aylesbury or Pekin; Appleyards, Saxony, Muscovies, and all of the Mediumweight, Lightweight, and Bantam breeds are naturally lower in body fat. Then, rather than pushing ducklings for the fastest possible growth, they should be allowed to forage for a portion of their food. The birds should be given just enough feed to keep them healthy and growing well. When butchered at 12 to 20 weeks of age, the fat content of birds raised in this manner is similar to that of wild ducks. In a trial involving 30 Rouen drakes from a production-bred strain, we found that when this method was employed, even these large meat birds produced carcasses that were no more fatty than roasting chickens.

Typical Growth Rate, Feed Consumption, and Feed Conversion of Pekin Ducklings Raised in Small Flocks

TREATMENT	AGE (WEEKS)	LIVE WEIGHT (LBS)	FEED PER BIRD (LBS)	FEED PER LB OF BIRD (LBS)
Fed free choice on a 20% protein, 1,400 kcal/lb ration from 0–2 weeks; then fed 16% protein, 1,400 kcal/lb ration from 3–12 weeks. Exposed to 24 hours of light daily and raised in confinement.	5	5.07	10.85	2.14
	6	6.10	14.52	2.38
	7	6.85	18.43	2.69
	8	7.43	22.36	3.01
	9	7.87	26.52	3.37
	10	8.36	31.43	3.76
	11	8.58	36.47	4.25
	12	8.43	40.55	4.81

Sexing Ducklings

Over the years, I have been told many secret methods for determining the sex of day-old ducklings. When tested, many of these techniques have proven to be only 50 percent accurate at best. To my knowledge, the following methods are the most trustworthy procedures for sexing live ducklings.

Vent Sexing

The *only* sure way to sex ducklings of all breeds before they are 6 to 8 weeks old is by examining the cloaca. While it is much easier to vent sex waterfowl than land fowl, this procedure still requires practice, an understanding of the bird's physiology, and care to avoid permanent injury to the bird. Too often, persons attempt to sex ducklings without first acquiring the needed skills, injuring the birds or wrongly identifying the gender.

At best, a written account on sexing ducklings is a poor substitute for a live demonstration by an experienced sexer. I *highly recommend* that you have a knowledgeable waterfowl breeder show you how to vent sex ducks before trying it yourself.

While ducks of all ages can be sexed by this method, I suggest that novices practice on birds that are 2 to 3 weeks old. At this age, ducklings are easily held and their sex organs are large enough to be readily identified. As birds reach maturity, the sphincter muscles that surround the vent become stronger, making it more difficult to expose the cloaca. Extremely young ducklings are usually harder for the beginner to hold for sexing, and the sex organs are tough to identify until a person knows what to look for. There are three points to keep in mind when vent sexing ducklings:

1. Young ducklings are *extremely* tender and clumsy fingers can kill the bird or permanently injure them.

2. A bird has not been sexed *until* the cloaca has been exposed. People often assume that if they can't find a penis after applying a little pressure to the sides of the vent, the bird is a hen. However, a duckling cannot be identified as a hen until the cloaca has been inverted and no penis is evident.

3. The sex organs of ducklings are tiny, so it is *essential* that birds be sexed outside on a sunny day or under a bright light.

You should be able to sex ducklings if the following steps are carefully implemented. There are several methods for holding birds for sexing, but I'll describe only the way I find most comfortable.

Step 1. Hold the bird upside down with its head pointed toward you. If the duck is large, its neck and head can be held between your legs.

Step 2. Use the middle and/or index finger of your right hand to bend the tail to a position where it is almost touching the bird's back.

Step 3. Push the fuzz or feathers surrounding the vent out of the way so you can see what you are doing.

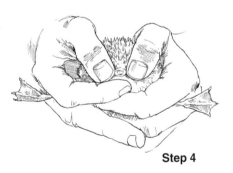

Step 4

Step 4. Place your thumbs on either side of the vent and apply pressure down and slightly out.

Step 5. Use your index finger of the right hand to apply pressure down and out on the backside of the vent. This inverts the cloaca and exposes the sex organs.

Step 5

Vent Sexing the Small Duckling

1. Vent sex young ducklings under good lighting. Hold the bird upside down with its head pointed toward you and double its tail back with an index finger.

2. With the legs out of the way, position your thumbs on opposite edges of vent and place free index finger just behind the vent. To invert the cloaca, simultaneously apply pressure down and out with thumbs and index finger.

3. Almost transparent, the penis of a week-old drake is visible in the center of the inverted cloaca. A duckling cannot be identified as a hen until the cloaca is inverted as shown and no penis is present.

If It's a Drake

If the duckling you are sexing is a male, a corkscrew-shaped penis will pop up near the center of the cloaca. (You must look carefully, since sometimes only the tip of the penis will be visible.) In drakes only a few days old, the penis is extremely small and almost transparent. By the time males are a couple weeks of age, their sex organs are easier to see because of their larger size and deeper white or yellowish color.

If It's a Hen

Should you have a hen, no penis will be visible when the cloaca is inverted. In hens that are more than several weeks old, the female genital eminence is often visible as a small, dark (usually gray) protuberance that resembles an undeveloped penis.

If You Have Difficulty

The most common problem encountered by inexperienced sexers is getting the cloaca inverted. If this is your problem, make sure that:

- ◆ You have the bird's tail bent back double as in the illustration.
- ◆ You are applying pressure down and out simultaneously with both your thumbs and the right hand index finger.

As with most skills, your speed and accuracy in sexing will improve with practice. If you get discouraged on your initial attempts, wait a few days until you have regained your confidence and then try again.

Vent Sexing the Older Bird

1. A convenient method for restraining large ducks for sexing is to turn the bird upside down and hold its neck between your legs.

2. After bending the tail back (A), push aside the feathers surrounding the vent, position your thumbs on the sides and your index finger on the back of the vent, and invert the cloaca by applying pressure down and out.

3. The cloaca and penis of an 8-week-old drake (B). If the cloaca is inverted sufficiently to expose two dimples and still no penis is evident, it can be assumed that the bird is a hen.

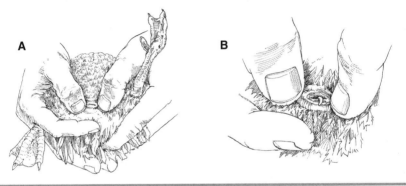

Sexing by Voice

By 5 to 8 weeks of age, hens of most breeds can be distinguished by their voices. To sex birds by this method, catch each duckling individually, and as it protests its predicament, listen carefully for the hen's distinctive harsh quack. The young male's vocalization is intermediate between its baby and adult voices, resembling an elongated "wongh." Due to the structure of the voice box, drakes are incapable of producing the hard quack sounds of the female.

Sexing by Bill Color

It is possible to sex purebred ducklings of some varieties by their bill color as early as 6 weeks of age. In gray varieties such as Calls, Mallards, and Rouens, the bill of a drake normally turns dull green while a hen's becomes dark brown and orange. The hen in most other colored varieties has a darker bill than the drake. This method is unreliable with some crossbred ducklings as variation in bill color can be a sign of genetic differences.

Sexing by Plumage

Drakes do not acquire their adult nuptial plumage until they are 4 to 5 months old. However, in many colored varieties, there are sufficient variations in the juvenile feathering of drakes and hens that make it possible to differentiate the sexes of ducklings that are 5 to 8 weeks old. In gray varieties such as Calls, Mallards, and Rouens, the feathers of the crown of the head and back often are darker and have less brown penciling in drakes than in hens.

Marking Ducklings

Several methods can be employed to mark individual birds, including the use of leg or wing bands, and the notching and perforating of the webbing between the duckling's toes.

Bands and toe punches are available from poultry equipment suppliers and feed stores. When leg bands are applied, care must be taken to use rings that are large enough to permit free circulation of blood to the feet.

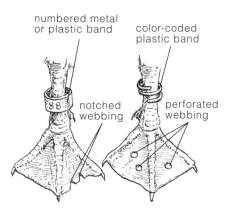

numbered metal or plastic band

color-coded plastic band

notched webbing

perforated webbing

Marking birds for easy identification

A Remarkable Rescue

One morning, my wife, Millie, found a 12-day-old duckling wedged in the lip of a water fountain. The only sign that the soaked bird was still alive was a barely perceptible movement of a leg. Millie put the bird in a small bucket of lukewarm water and carefully held it there with its head above the surface. In five minutes, the little fellow was able to hold its head up. Millie toweled him off and then, using a hand-held hair dryer set on low, dried his down, being careful not to overheat him. We placed the duckling in a 90°F battery brooder with a group of 5-day-old ducklings. An hour later, the "miracle" duckling was up and running around with no sign of his near-death experience.

Brooding Ducklings with Other Birds

Many of us who have duck flocks also raise other poultry. Consequently sometimes we're brooding ducklings and other birds at the same time. To save space and equipment, it is tempting to brood the various species together.

However, the young of ducks, geese, quail, guineas, chickens, and turkeys have unique habits, temperaments, growth rates, and management requirements that can cause serious problems when birds are raised in mixed groups. While it is not impossible to brood ducklings with other young poultry — goslings and ducklings get along fairly well — many problems are avoided and each grows better if they are brooded separately.

Perennial ryegrass turf, a water fountain, and plenty of shade from a row of hardy bamboo make this an excellent environment for these 8-week-old waterfowl. Raising a few geese with ducks helps reduce losses to predators.

FOURTEEN

Managing Adult Ducks

Adult ducks do not require a lot of time-consuming care or specialized facilities. By using common sense and being observant, novices can manage ducks successfully.

Basic Guidelines

The most common error made with ducks is trying to raise them like chickens. For good results, every species of fowl must be managed somewhat differently, and the following are absolute musts in duck care:

1. **Protect your birds from predators.** Ducks raised in small flocks seldom die from disease or exposure to severe weather, but quite a number are lost to predators. In most locations, ducks should be penned in a building or a tightly fenced yard every night.

2. **Supply a balanced diet in an adequate quantity.** Following close behind predators, improper nutrition is the second leading cause of problems encountered by duck keepers.

3. **Provide a steady supply of drinking water.** Ducks do not thrive if they are frequently left without water.

4. **Furnish suitable living conditions.** Keeping ducks locked up in yards covered with deep mud and stagnant waterholes is an invitation for trouble.

5. **Do not disturb them more than necessary.** Waterfowl thrive on tranquility.

6. **When catching, do not grab or carry them by their legs.** The legs and feet of ducks are easily hurt.

Housing

Ducks require minimal housing. Unlike chickens, they prefer to stay outside day and night in most weather. In mild climates it is possible to raise ducks without artificial shelters. A windbreak made of straw bales will provide sufficient protection to keep ducks comfortable in regions where temperatures occasionally dip below freezing. A more substantial shelter is needed in areas where extremely low temperatures are common.

Because waterfowl normally perch on water or the ground at night, the main reason for housing ducks in mild weather is to protect them from night-wandering predators (see appendix C, Predators, page 291). A tight fence at least 4 feet high enclosing the yard is enough to stop many predators. However, in areas where thieves such as weasels, mink, raccoons, owls, and wild or domestic cats are common, it is much safer to enclose ducks in a varmint-proof building or covered pen at nightfall.

If you do not have an empty building that can be converted into a duckhouse, an inexpensive shedlike structure can be constructed. In the illustration on page 193, a practical duckhouse is shown. This type of shelter is portable and can be used for either ducklings or adult birds.

Masked Bandits

Their clownish countenance notwithstanding, raccoons are the most destructive of all fowl thieves in many regions of North America. Adaptable to both rural and urban settings, smart, persistent, strong, amazingly dexterous, and able swimmers, this large member of the weasel family is aptly labeled "super predator."

Raccoons will climb up and over the tallest fence as if it were a ladder put there for their convenience. And with their nimble paws, they have been known to reach through wire netting with 1-inch-wide openings, grab dozing birds, and eat them through the fence.

The only sure way to keep ducks safe from these nocturnal hunters is to lock the waterfowl *every* night in a building or covered pen (one that has no opening larger than a half inch in diameter in the bottom 30 inches of the walls) or a tightly fenced yard that is protected by properly installed electrical fencing. (See page 196 for fencing ideas, and appendix C, page 291, for additional information on predators.)

The Triplex Duck Run

Where we live in Western Oregon, we have long, wet winters, dry summers, and numerous predators — most notably raccoons, opossum, coyotes, foxes, bobcats, cougars, eagles, and large owls. If ducks are left outside at night here, sooner or later they will become some critter's dinner. Furthermore, with all the ducks we have, they would demolish grassy runs if allowed outside during the wettest times of the year. Over the years we have found that the triplex duck run ensures safety for the birds, provides them a healthy environment, and protects the grass.

The triplex duck runs consist of three main components:

1. An area inside a building that can be tightly closed at night against all predators and that provides a minimum of 2 to 6 square feet per bird depending on their size and temperament. This area is bedded with coarse cedar sawdust.

2. A bedded outside yard providing an additional 2 to 10 square feet per bird, that ducks have access to during the day. To prevent mud, we cover these yards with 3 to 4 inches of coarse, round gravel, topped with 2 inches of sand, covered with a layer of coarse cedar sawdust. To reduce the amount of maintenance to keep the yards in good condition, over the years we have covered most of them with roofs.

3. A grassy yard that is irrigated during the summer is available to the ducks during nice weather. Shade is supplied by large shrubs, trees, and grape arbors. We try to supply a minimum of 50 square feet per bird.

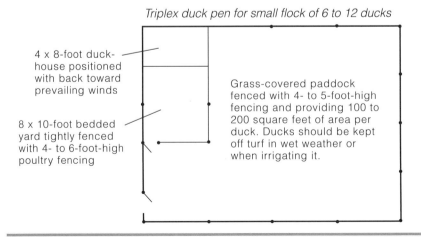

Triplex duck pen for small flock of 6 to 12 ducks

4 x 8-foot duck-house positioned with back toward prevailing winds

8 x 10-foot bedded yard tightly fenced with 4- to 6-foot-high poultry fencing

Grass-covered paddock fenced with 4- to 5-foot-high fencing and providing 100 to 200 square feet of area per duck. Ducks should be kept off turf in wet weather or when irrigating it.

Since ducks are short, there is no need to have the walls of the duck-house more than 3 feet high at the lowest point, except for your convenience in gathering eggs and cleaning out old litter. By having the nests attached along the outside and outfitted with a hinged top, eggs can be gathered without entering the shed. Three solid walls and a wire front are recommended except in regions with severe winters, where it is better to have a closed front. Good ventilation is essential, however, even in cold climates, because ducks will fare poorly if they are forced to stay in stuffy, damp quarters.

Dirt, sand, wood, or cement floors can be satisfactory in duck buildings. To keep predators from burrowing into buildings with dirt or sand floors, it may be necessary to place a wire, wooden, or cement barrier around the outside perimeter of the structure.

To ensure good drainage, place the shed on a slope or build the floor up 4 to 8 inches above ground level with dirt, sand, or bedding. If the duckhouse is located in a low spot, water will accumulate, making it impossible to keep the litter dry.

When medium- to large-sized ducks are housed only at night, allow at least 2.5 to 4 square feet of floor space per bird. If you anticipate keeping your ducks inside continuously during severe weather, a minimum of 6 to 8 square feet of floor space per bird should be provided. Bantam breeds require approximately half the floor space of larger ducks.

The attached nests on this practical duckhouse make egg-gathering easier.

Cold-Weather Housing

Ducks are well prepared to remain comfortable in freezing weather with their thick garb of feathers and down. Nonetheless, feed consumption can be reduced and egg production increased if ducks are protected from the severity of bitter northern winters, particularly at night when the birds are inactive.

Insulated Quarters

Insulating a small duckhouse is inexpensive and can mean the difference between few or many eggs during the long winter months. If the duckhouse has double-wall construction, air spaces can be filled with sawdust, chopped straw, or a commercially prepared material. Another satisfactory way to insulate shelters is to stack straw bales against the outside walls from ground level to the roof.

Deep Litter

A thick layer of bedding in the duckhouse keeps ducks off cold floors and reduces the penetration of cold from the ground during periods of low temperatures. The deep litter system actually goes one step further by producing heat through decomposition of the litter and manure. This system has been tested in cold climates and has been found to increase winter egg production by up to 20 percent. It also eliminates the need to clean out duckhouses more than twice yearly, and produces excellent organic matter for the home garden.

To employ the deep litter method, follow these three steps: (1) cover the duckhouse floor with 4 to 6 inches of dry bedding; (2) periodically stir the bedding and add fresh material to keep the floor in good condition; and (3) clean out the old litter each spring. Because the litter will accumulate to a depth of 18 to 30 inches, the walls of duckhouses where this system is used need to be at least 4 feet high.

The Duck Yard

Most serious duck raisers have a fenced yard where birds are locked in at nighttime, or continuously if space is limited. The ideal yard has a minimum of 10 to 25 square feet of ground space per duck, natural or artificial shade, a slope that provides good drainage, and a surface that is free of stagnant waterholes and deep mud.

Raising Ducks with Chickens

When not overcrowded, ducks can be raised with chickens and other species of birds. However, because ducks have wet habits and wash their bills and faces in drinking water, it is best to have separate drinking water for the chickens on an elevated perch out of the reach of ducks. Also, there are diseases that different species of birds can carry and spread to one another. Commercially, ducks and chickens (and other landfowl) should be kept separate.

If the soil on the land where you live has poor drainage, the duck yard should be covered with pea gravel, sand, straw, or wood shavings, with the center of the yard built up higher than the outside edges. When choosing a location for your duck yard, keep in mind that a dense population of birds can kill young trees due to the high nitrogen content of duck manure. The area around the perimeter of yards should be kept free of briar patches, low-growing bushes and trees, and tall grass or other hiding places for predators. Around our duck yards, we keep a swath at least 25 feet wide mowed or grazed low to discourage surprise attacks by fowl-loving meat-eaters. It is fine to have trees or shrubs in duck yards for shade and the edible fruit or seeds they produce, but remember that woody plants that grow on the outside and overhang fences provide ready access for climbing predators.

Keeping Ducks on Wire Floors

Adult ducks can be kept in houses or hutches with wire-covered floors or in elevated, all-wire cages. The main advantage of this system is that birds can be maintained in extremely limited space without problems of sloppy litter or muddy yards. Some disadvantages are that feed consumption is higher since the birds cannot forage, foot problems can develop (particularly among the larger breeds), and life is less enjoyable for the ducks.

To prevent lameness, wire mesh used on floors should not have openings larger than 1 inch by ½ inch. However, if finer mesh is used, manure may not pass through, causing buildup problems. A minimum of 2 square feet of floor space per medium- or large-breed duck should be allowed, with double or triple this amount being recommended when feasible.

Choosing Fencing Material

The main purpose of fences is to keep ducks in and predators out. Most breeds of ducks can be kept in with a 2- to 3-foot-high barrier if the openings are of an appropriate size and the bottom of the fence is close to the ground.

However, to keep dogs and coyotes out, tight fences at least 4 to 5 feet tall are necessary. (In general, dogs and coyotes will not jump over a 5-foot fence into unfamiliar territory, although there is no guarantee they never will.) Our farm is surrounded by coyotes, but we have never had one breach our 5-foot-high perimeter fences. Remember: Dogs, coyotes, and ducks prefer to go under or through a fence rather than over its top.

The following is a brief review of some types of fencing material:

1. Wire chicken netting. This lightweight material commonly comes with either 1- or 2-inch openings and is relatively inexpensive. Raccoons have been known to tear through it, posts need to be used every 4 to 6 feet to keep the bottom tight, and ducks will sometimes get their bills snagged in it.

2. Deer and rabbit woven fencing. A moderately heavy-duty fencing with vertical stays that are 6 inches apart and horizontal wires that start out 1 inch apart, gradually increasing to 4 inches apart at the top. It is generally available in heights from 4 to 8 feet. Properly installed, this wire will stay tight for decades. It is what we use on our farm for perimeter fences.

3. Woven field fencing. A moderately heavy-duty fencing usually used for sheep, hogs, and cattle. It has vertical stays that are either 6 or 12 inches apart and horizontal wires that start out at 2½ to 3 inches apart, gradually increasing to 5 inches or more at the top. It will last a long time, but all breeds of duck except for the larger ones can walk or squeeze through the openings.

4. Welded wire. A strong, rather rigid wire that comes with openings of various dimensions, including ½ x 1 inch, 1 x 1, 1 x 2, and 2 x 4. Heights range from 2 to 8 feet. We use ½ x 1 inch and 1 x 1 meshes for floors of water porches and water platforms (areas where the birds drink), and 2-foot-high 1 x 2-inch mesh for divider fences in breeding and brooding yards. (These short fences are easy for humans to step over.) In general, we do not use wire with 2 x 4-inch openings because ducks will occasionally get their heads and wings caught in them.

5. Chain-link fencing. A heavy-duty, very strong wire and expensive product that lasts a lifetime under many applications. It comes in many heights and is excellent for strong perimeter fences if installed properly.

6. Electric fencing. When installed properly in conjunction with woven wire, electric fences can keep out climbing predators such as raccoons. Typically, it is installed with two or three strands of horizontal electric wire placed 2 to 6 inches on the outside of the woven-wire fencing. One strand runs 4 to 6 inches above the ground, a second runs 18 to 24 inches above the ground, and a third runs near the top of the fence.

7. Electroplastic fencing. As long as large predators do not "bulldoze" into this lightweight electric netting, it does a great job of keeping birds in and predators out. It comes in heights ranging from 22 to 42 inches and can be moved readily. Electroplastic fencing can be a lifesaver when installed around a setting duck that has stationed its nest outside of a protective yard or building. (See appendix G for sources.)

8. Gamebird netting. This netting comes in rolls of different widths and is excellent for covering pens and yards where winged predators such as owls and eagles are problematic. (See appendix G for sources.)

Bedding

Ducks can withstand wet weather, but they should *never* be forced to stay in muddy, filthy yards or buildings. The floors of buildings should be covered with several inches of mold- and toxin-free litter such as sand, sawdust, wood shavings, peanut hulls, peat moss, crushed corncobs, flax, or straw. In muddy or snow-covered yards, mounds of bedding 4 to 6 inches thick should be provided to give ample space where all ducks can roost comfortably. Add new layers of clean bedding as needed.

Shade

Ducks must have access to shade in hot weather. They suffer if forced to remain in the sun at temperatures above 70°F. When ducks are confined to a yard lacking natural shade, a simple shelter should be provided that will supply adequate shade for all birds residing there. Feed troughs can also be located under this cover to protect the feed from exposure to the elements.

A simple shelter provides shade and protects feed from sun and rain.

Nests

Clean eggs hatch better and retain their freshness for eating longer than eggs that must be washed. Having adequate nests for your hens will help produce clean eggs and reduce the chances of eggs being broken. To allow hens to become familiar with them, install nests 2 weeks before you expect the first eggs. One nest for every four or five hens is sufficient.

For medium to large hens, nests approximately a foot square and 12 inches high are recommended. Duck nests are placed on the ground and do not need solid bottoms. If nests are inside buildings, they do not require a top. Covering the nest bottoms with burlap aids in the production of clean eggs. Keeping nests well furnished with clean, dry nesting material such as sawdust or straw will encourage hens to use them and result in fewer broken or soiled eggs.

Portable Panels

Lightweight panels constructed from 1 inch × 2 inch or 1 inch × 4 inch lumber and chicken netting frequently come in handy. They can be used to divide pens, for isolating hens that are setting or have young, as a catch pen in a large yard, and for small pens that can be moved around to utilize grass in different areas of the home plate. To retain most breeds of ducks, the panels only need to be 2 or 3 feet high. Panels that will be used to pen day-old ducklings should be covered with 1 inch × 1 inch chicken netting.

Farm and ranch supply stores often sell galvanized stock panels that are 16 feet long. We use dozens of these handy panels around our farm. The 34-inch-tall "hog" version is handy for making temporary pens for young duck stock or breeders. The 52-inch-high style (choose panels with small openings at the bottom so ducks cannot walk though them) will keep out most dogs and can be used for gates or around the perimeter of duck yards.

Swimming Water

Ducks enjoy swimming, and add charm and interest to lakes, ponds, and brooks. Where natural bodies of water do not exist, earthen or cement ponds can be constructed. Be sure to design them for easy drainage and cleaning. To keep earthen banks from eroding, rip-rap (large rocks) is recommended. A 4-inch-deep layer of sand or gravel around the perimeter of a pond is helpful in keeping ducks from tracking in mud and drilling with their bills.

Under many circumstances, rather than digging a pond, using children's wading pools, masonry mortar pans (available at hardware stores), or large dishpans has many advantages. They are easy to empty, they can be moved to fresh ground every day to prevent mud, or they can be placed on top of a wire platform or a bed of gravel.

Contrary to popular belief, ducks can be raised successfully without water for swimming. In fact, there are some advantages to having only drinking water. If a small pond is overcrowded, sanitation problems often arise. Ducks must *not* be allowed to swim in or drink filthy, stagnant water.

Our Breeding Schedule

Because we are working with many breeds and strains of ducks, we use a large number of relatively small breeding pens. Matings consist of pairs, trios, pens (1 drake: 3–5 ducks), and flocks containing a ratio of 1 male: 4–6 ducks. Our breeding season is January to June; therefore, our schedule is designed to maximize production during these months.

1. Prior to making up the breeding pens, the ducks are blood tested for pullorum-typhoid.

2. Four weeks before hatching eggs are saved, ducks are placed in their breeding pens and artificial lights are used to lengthen the days to 14½ hours of light daily.

3. We provide one nest box for every one to four females inside the duck barns. To encourage exercise, feeders are located at the far end of the pens inside the duck barns while the waterers are located 15 to 30 feet away in the outside runs.

4. Three weeks before hatching eggs are desired, ducks are switched (over the course of several days) from a 14-percent-protein maintenance ration to our 19-percent-protein breeder pellets.

5. For large breeds that may have reduced fertility, we provide water pans, which stimulate mating in more lethargic birds.

6. During wet weather, the ducks have access to inside shelters and covered runs. When the grassy yards are sufficiently dry so that the ducks will not muck them up, they are allowed outside from 10 A.M. until evening.

7. To protect them from predators, they are locked up every night in the duck barns.

8. As soon as the laying season is over, the birds are switched from the breeder ration to a maintenance ration with 0.6 to 1 percent calcium and 14 percent protein.

Drinking Water

Ducks must have a *steady* supply of reasonably clean drinking water. So the birds can wash out their nostrils, waterers should be designed to provide a water source that is at least 3 inches deep. Buckets, dishpans, and hot water tanks cut in half vertically make satisfactory drinking containers for adult ducks. For a pair or two of birds, a 1-gallon tin can will suffice. Whatever drinking container you choose, make sure it is not one where ducks could go into the container and not be able to easily get out.

Because ducks frequently wash their bills, drinking water would need to be changed many times daily to be kept clear. While such repeated changing is not necessary, the water should be replaced several times weekly and must never be allowed to become putrid.

To keep unhealthful mudholes from developing around watering areas, water fountains should be placed on wire-covered platforms. (A simple method is to place a 3- or 4-foot-square piece of welded wire mesh — with 1-inch-square or ½ x 1-inch openings — on the ground underneath all waterers.) For additional protection, a pit 12 to 24 inches deep can be made underneath the platform and filled with gravel.

During cold weather when there is soft snow, ducks can sustain themselves for a time by eating the icy crystals. However, it is preferable to give them warm drinking water twice daily when the temperature is cold enough to freeze all available liquid.

If ducks have been conditioned to not have drinking water at night at least 6 weeks prior to onset of lay, egg production is minimally affected by withholding drinking water for up to 9 hours each night. *For this procedure to work, consistency is paramount!*

Raising Ducks on Salt Water

Ducks can be raised successfully in marine areas. Most ocean bays and inlets are teeming with an abundant supply of plant and animal life that ducks relish. Because domestic ducks have a lower tolerance for salt than do wild sea ducks, sweet drinking water should be supplied at all times. Birds that are raised for meat on marine waterways need to be confined in a pen or yard and fed a grain-based diet for 2 to 4 weeks prior to butchering to avoid fishy-flavored meat.

Nutrition

Feed is the single most expensive item in raising ducks, normally representing 60 to 80 percent of the total cost. Finding ways to save on feed expenditures will significantly decrease the cost of your duck project.

For top egg production, ducks must be fed an adequate amount of concentrated feed having 16 to 20 percent crude protein. Three weeks before the first eggs are expected and throughout the laying season, a laying ration should be fed free-choice or twice daily. (See the table on the facing page for recommended daily feed allowances.) Duck hens lay the best when they are in a semi-fat condition, but excessive fatness is harmful and must be avoided. A sudden change in the diet of hens that are laying normally results in a sharp decline in egg production, throwing hens into a premature molt from which it will take 6 to 10 weeks to recover.

While not in production, mature ducks can be given a maintenance ration containing 13 to 14 percent crude protein, and fed just enough to keep them in good condition. The quantity of feed required by birds is highly affected by the weather. During cold periods, ducks must eat considerably more feed than when temperatures are moderate.

Fish by-products such as fish meal are excellent sources of high-quality proteins, and are used in many commercially prepared poultry feeds. These ingredients are excellent for immature birds and adult breeders that produce hatching eggs, but laying hens that supply eating eggs must be given feeds

Common Causes of Poor Egg Production

1. Sudden change in appearance, texture, flavor, or content of feed
2. Birds unexpectedly left without feed or water
3. Sudden or periodic changes in hours of daily light exposure
4. Decreasing length of daylight
5. Periodically being startled
6. Improper diet
7. Heavy infestation of internal and/or external parasites
8. Overcrowding
9. Filthy living quarters
10. Contaminated water

that are low in (less than 4 percent), or do not contain, fish products to pre-vent the production of strong-flavored eggs.

Feeding Programs

In formulating a feed program for your adult duck flock, try to find ways to obtain a high degree of productivity while using food resources that normally go unharvested.

Natural

The least expensive way to feed ducks is to make them forage for their food. This option is limited to situations where wild and natural foods are in abundant supply and when top egg production is not important. It must be remembered that the quantity of natural feed fluctuates widely during the various seasons of the year. While there may be periods when ducks can find most and possibly all of their own feed, there will probably be times when most of their food will need to be supplied in the form of grain and mixed rations. Ducks must *never* be allowed to deteriorate and become thin due to lack of feed. Negligence in this area can permanently damage their productivity.

Suggested Feeding Schedule for Adult Ducks

Size of Duck	Holding Period When Birds Are Not Producing Lbs. of 12–14% Protein Feed*	3 Weeks Prior to and During Laying Season Lbs. of 16% Protein Feed*	2 Weeks Prior to Butchering Mature Ducks Lbs. of High-Energy Feed*
2–3 lbs.	0.15–0.25	0.20–0.30	Free choice
4–5 lbs.	0.20–0.30	0.30–0.40	Free choice
6–7 lbs.	0.25–0.35	0.40–0.50	Free choice
8–9 lbs.	0.30–0.45	0.45–0.60	Free choice
Muscovy	0.30–0.40	0.40–0.60	Free choice

*per bird daily

The quantities of feed that you need to supply to ducks are highly dependent on the availability of natural foods, the climate conditions, and the quality of feed (e.g., birds require larger amounts of high-fiber foods than low-fiber foods to meet their energy requirements).

Grains

Whole grains, by *themselves*, are not a complete poultry feed. However, if ducks are given a protein concentrate or can forage in waterways or pastures, a supplement of grain will often satisfy their dietary needs, particularly when the birds are not in production.

Ducks are not as fond of oats and barley as they are of other grains. But these two grains are good waterfowl feeds and the birds will learn to eat them if made to do so.

Corn is a high-energy grain and an excellent feed for cold weather and for fattening poultry — if you desire fat meat. During hot weather, the diet of ducks should not consist of more than 60 to 80 percent corn. If waterfowl are fed too much corn during exceptionally hot weather, egg production drops and health problems can arise.

Home-Mixed

It may be practical to mix your own feed if the various ingredients are available at a reasonable price. Mixing a home ration for ducks is simpler than for chickens. All ingredients for chicken feeds need to be finely ground; otherwise the birds will pick out the grains and leave the finer particles. This problem is usually not encountered with ducks since they tend to scoop up their feed rather than picking it up one kernel at a time.

The formulas in the table in appendix A, Home-Mixed Rations for Adult Ducks, are examples of the types of feeds that can be mixed at home for small flocks of ducks. If a commercially prepared vitamin:mineral premix is used, the rations provided in the table Complete Rations for Adult Ducks (Pelleted) can be prepared at home as well. The suggested ingredients in the tables can be substituted with similar products that are more readily available in your area.

Commercial

The simplest way to provide ducks with a balanced diet and stimulate top production is to purchase commercially mixed concentrated feed. These rations are formulated to ensure a proper balance of carbohydrates, fats, proteins, vitamins, minerals, and fiber.

In many areas, feeds manufactured specifically for breeding and laying ducks are not available. Nonmedicated turkey and chicken feeds can be used in place of duck mixtures. (Chicken rations usually need to be supplemented with additional niacin. See Niacin Requirements on page 180.) Ducks waste less feed if they are fed *pellets* rather than crumbles or mashes.

If you are raising a large number of ducks, it may be economical to have a local feed mill custom-mix your duck feed. In one of the tables in appendix A, Complete Rations for Adult Ducks (Pelleted), feed formulas that we have used successfully for a number of years are given.

Because wheat is our most abundant and cheapest grain here in the Northwest, we use it as the base for our rations. In your area, other grains may be lower-priced and they should be used. Your grain dealer can help you formulate a feed that best utilizes available resources.

Growing Duck Feed

It is possible to raise a good portion of the feed required by a flock of half a dozen ducks on a 50 foot x 50 foot plot of land. Field corn yields large quantities of ears that are easily harvested and shelled by hand or broken in two and thrown to the ducks to shell for themselves. Grains such as wheat and rye provide fall and spring grazing and, when the seeds are mature, the ducks can be allowed into the patch for a short time daily to harvest their own feed. Grain sorghums such as kafir and milo produce large seed heads that can be hand-harvested and stored whole.

Our favorite perennial plants for producing significant quantities of food and shade in our duck yards include mulberries, Siberian pea shrub (*Caragena arborescens*), and grapes. (The grapes are grown on a rectangular four-posted arbor that is 8 feet tall.) The fruit and seeds from these plants require no harvesting because they fall to the ground when mature and the ducks eagerly consume them.

A large assortment of natural crops is available for planting in and around bodies of water. Some of the favorites of ducks include wild celery, wild rice, wild millet, small bulrush, smartweed, and chufa tubers. Addresses of nurseries specializing in natural game bird crops can be found in hunting and outdoor magazines. Ducks also enjoy and benefit from vegetable and fruit produce that cannot be used in the kitchen. See the chart on page 287 on complete rations for adult ducks.

Pasture

While ducks can be raised in barren yards or pens, they enjoy succulent vegetation when it is available. Good-quality forage lowers their feed

consumption by approximately 10 percent and lessens the possibility of vitamin deficiencies. Ducks cannot eat mature pasture (except the seeds), so it
should be mowed occasionally to encourage new growth.

Orchards are an excellent location for duck pastures. The vegetation
smothers out noxious weeds and provides a protective and attractive ground
cover. Ducks significantly reduce the numbers of harmful insects in orchards
and clean up diseased and windfall fruits.

There are a variety of grasses and legumes that make good permanent
pasture for ducks. The main requirements are that the plant thrives in your
region and that it produces succulent forage that your duck finds palatable.
A mixture of Perennial Rye grass and one of the clovers such as New Zealand
White, Lodino, or White Dutch will provide good grazing in many localities.
Your local agriculture extension service specialist can give advice on what
varieties do well in your climate.

Feeding Insects to Ducks

Insects are a bountiful source of high-quality protein. While ducks of all
ages are accomplished bug hunters, their consumption of winged insects can
be increased by burning a low-wattage bulb 12 to 18 inches above ground
level in the duck yard at night. As the insects swarm around the light, the
birds have hours of leisurely dining.

Using Surplus Eggs and Milk

Eggs and milk are excellent sources of nutrition for poultry. Being "complete" foods, they are particularly valuable in home-mixed rations to help
ensure that birds are receiving a balanced diet. In general, when eggs and/or
milk are added to the diet of ducks, they grow better, have glossier plumage,
and lay more eggs.

The only practical way to feed liquid milk to ducks is to mix small quantities into a dry ration. When milk is given in pans or buckets, ducks play in
it, covering themselves and the surrounding area with the sticky stuff. Under
these conditions, food poisoning and eye infections often develop.

Eggs should *always* be hard-boiled before they are given to poultry.
Feeding raw eggs can result in a biotin deficiency and often leads to birds
eating their own eggs. To avoid spoilage, milk and eggs should not be mixed
with the duck ration until feeding time.

Feeding Leftovers

Kitchen and garden refuse must be fed to producing birds in moderation. Leftovers normally are high in starch, fiber, and liquid and low in most other nutrients. Hens cannot lay well if their minimum requirements for protein, minerals, and vitamins are not being ingested.

Grit and Calcium

To digest their feed to the best advantage, ducks need to have a continuous supply of granite grit, coarse sand, or small gravel. Four weeks prior to and throughout the laying season, hens need to have their diets supplemented with calcium if they are going to lay strong-shelled eggs. Most manufactured laying feeds contain the correct amount of calcium, but when grains or home mixes are used, a calcium-rich product such as dried, crushed egg shells, oyster shells, or ground limestone should be fed free-choice.

Feeders

When adult ducks are fed a limited quantity of feed, sufficient feeder space must be allowed so that *all* the birds can eat at one time without crowding. Otherwise, less aggressive and smaller ducks may be pushed aside and not get their full share.

Durable Containers

Flexible rubber pans (with sides approximately 4 inches high) are commonly available in various sizes at feed and farm supply stores. These receptacles do not crack when they freeze and make excellent feeders and water/bathing pans that last for years under normal use.

Approximately 4 to 6 linear inches of feeder space should be allowed per bird. Ducks can eat from both sides of most feeders, so a trough 5 feet long provides 10 linear feet of feeder space, enough for a flock of 20 to 30 birds. V-shaped troughs, with spinners mounted along the top to keep birds out, are easily constructed and provide sanitary eating conditions.

Finishing Roasting Ducks

When ducks have been forced to forage for much of their feed during the growing period, they can be fed grain or finishing pellets free-choice for 2 or 3 weeks prior to butchering. If they have ranged widely, it is advantageous to pen them in a restricted yard (allowing a minimum of 10 to 25 square feet per bird) during the finishing period. This time of heavy feeding and restricted exercise will encourage the ducks to gain weight rapidly, resulting in a larger carcass and succulent, tender meat.

The Laying Flock

Most duck hens lay before 8 A.M. So that eggs are not lost in fields or bodies of water, it is a good practice to confine the laying flock to a yard or building at night.

Unless eggs are going to be used for hatching, there is no need to have drakes. Hens lay better without frequent mating activity.

Lighting Needs

When top egg production is desired during the short days of fall, winter, and early spring months, duck hens, like chickens, *must* be exposed to artificial lighting. A minimum of 14 hours of light daily is required to keep duck hens laying well. It is essential that the length of daylight *never* decreases while hens are producing, or the rate of lay may be severely diminished. Even a reduction of only 15 to 30 minutes per 24-hour period for several days can negatively affect heavily producing hens.

By using an automatic time switch that turns lights on before daybreak and off after nightfall, hens can be exposed to 14 to 16 hours of light daily.

To prevent premature egg production, the amount of light young hens are subjected to needs to be watched carefully. When pullets are exposed to excessive light or increasing day lengths between the ages of 12 and 22 weeks, they will begin to lay before their bodies are adequately mature. Early laying (before 18 to 20 weeks of age) can result in a shortened production life, smaller and fewer eggs, and greater possibility of complications such as prolapsed oviducts.

The intensity of light required to stimulate egg production is relatively low (approximately 1 foot-candle at ground level, which usually equals 1 bulb watt

Typical Effects of Management on Egg Production

	ANNUAL EGG PRODUCTION PER HEN		
TREATMENT	DOMESTIC MALLARDS	COMMERCIAL ROUENS	KHAKI CAMPBELLS
Fed whole or cracked grains; given access to pasture and pond; exposed to natural day length.	25–40	50–65	75–150
Fed 16% protein laying pellets; given access to pasture; exposed to natural day length.	50–75	75–100	175–225
Fed 16% protein laying pellets; given access to pasture; exposed to 16 hours light daily.	85–125	125–150	275–325

per 4 square feet of floor area when the bulb is at a height of 7 or 8 feet). When ducks are confined to a building or shed at night, one 40- to 60-watt incandescent bulb 6 to 8 feet above ground level will provide adequate illumination for each 150 to 250 square feet of floor space. In outside yards, one 100-watt bulb with a reflector per 400 square feet of ground space is recommended.

We use the following lighting schedule with good success for ducks that are hatched in March through July, and I recommend it if you desire maximum efficiency and production.

AGE	HOURS OF LIGHT DAILY
0–8 weeks	24 hours.
9–20 weeks	Natural day length.
21 weeks	Add 15 minutes of light weekly until 14 hours of total light (natural and artificial) is reached. If the day is naturally 14 hours or longer, add no artificial light. Once natural light falls below 14 hours, add enough light morning and evening to provide 14 hours. If egg production declines significantly, add 15 minutes of light weekly until a maximum of 16 to 17 hours of light daily is reached.
23 weeks	Add 15 minutes of light weekly until 15 or 16 hours of total light is reached. Maintain a constant level until hens stop laying or they are force-molted.

Identifying Producing Hens

Shortly before production commences, and throughout the laying season, the abdomens of hens swell noticeably. Hens that are in production can be identified by their large, moist vents and widely spread pubic bones. As the season progresses, the bright yellow bills of higher producing, white-plumaged ducks will fade to pale yellow. In ducks with colored plumage, the bill normally darkens during the laying season.

Encouraging Hens to Use Nests

To encourage ducks to lay where you want them to, place dummy eggs in nest boxes several weeks before the beginning of the laying season. While nest eggs can be purchased, I have found that turning them out on a lathe is an enjoyable rainy-day project. You do not have to worry about making them exactly the correct shape and size. Painting the wooden eggs makes them longer lasting and easier to clean. Any color will be accepted by the hens, but white or light-colored shades have the advantage of being more visible to birds.

Production Life of Hens

With good management, Lightweight Breed pullets begin laying at 17 to 24 weeks, while pullets of the larger breeds typically commence laying sometime from 20 to 30 weeks. Ducks lay the greatest number of eggs in their first year of production, but normally show only a minor decrease in productivity during the second and third year when eggs are larger in size. Good producers often lay some eggs until they are 5 to 8 years old.

Breaking Up Broody Hens

When ducks become broody, their egg production falls off or ceases. To induce broodies back into production, isolate them in a well-lit pen without nests or dark corners, and provide drinking water and feed. In these surroundings, hens normally lose their maternal desires in 3 to 6 days and can be returned to the flock. Because broodiness seems to be contagious, remove hens from the flock *promptly* when they show the first signs of wanting to set.

Identifying Problems in the Laying Flock

SYMPTOM	COMMON CAUSE
Thin- or soft-shelled eggs	Usually a vitamin D_3 or calcium deficiency; also high temperatures or abnormal reproductive organs.
Eggs decrease in size	A gradual decrease in size is common as the laying season progresses. However, hot weather, excessive environmental stresses, and dietary deficiencies accentuate the problem.
Odd-shaped eggs	Temporary malfunction of the reproductive organs; in some cases, abnormal oviducts.
Pale-yellow yolks	A diet lacking carotene, which is supplied by products such as yellow corn and green plants. Hens fed wheat-based rations without access to pasture or greens usually produce anemic-colored eggs.
Bright orange-red yolks	Diets high in corn and/or green feeds.
Blood or meat spots in egg interior	Internal hemorrhage. Eggs fit for eating; most people find these distasteful.
Blood on shell exterior	Ruptured blood vessel at the cloaca opening. Frequently occurs when young hens begin laying.
The back of head and neck bare of feathers; in extreme cases, skin is lacerated and scabby	Too many drakes, which results in excessive mating activity. Drakes sometimes have favorite hens.
Hens go into a premature molt	A sudden change in diet or lighting; birds frequently left without water; onset of hot weather.
Eggs lost in bodies of water, pasture, or hidden nests	Hens not locked in a pen or small yard at night, or are turned out before 8 or 9 A.M.

Force-Molting

Each year ducks lose their feathers and replace them with new ones. Hens normally do not molt while they are laying since their bodies cannot support the strain of growing feathers and forming eggs at the same time. Under natural conditions, this arrangement works out fine; the hen lays in the spring, hatches and broods her young until they are able to fend for themselves, and then she molts, with her new garb being ready for the cold winter months.

When ducks from prolific strains are managed for good egg yields, they typically commence laying at 17 to 24 weeks of age and lay well without long interruptions for 10 to 18 months. After this time, their feathers are worn and production normally drops substantially. Because it is practical to keep laying ducks for 3 years or more under some circumstances, it is common to force-molt them every 10 to 18 months to give their bodies a rest from laying and permit them to grow a new set of feathers.

The most common time to force-molt ducks is when egg production is down and temperatures are not bitterly cold. The time it takes ducks to molt and begin laying again varies, but is generally from 6 to 12 weeks.

Hens are thrown into a molt by sudden changes in their diet and environment. The following schedule is suggested as a simple method for force-molting the home duck flock.

1st day: Discontinue artificial lighting and remove all water and feed.

2nd day: Provide drinking water but no feed.

4th day: Commence feeding again, but substitute the laying ration with whole or rolled oats fed free-choice.

15th day: In addition to the oats, supply in a separate feeder .25 pounds per bird of a 16- to 18-percent-protein duck or waterfowl grower ration once daily — removing any of the grower that remains after 5 minutes.

42nd day: Gradually replace oats and grower feed with layer ration fed free-choice, and supply 14 hours of light daily. (If natural light is longer than 14 hours, add no additional light.)

Selecting Breeders

To maintain the productivity of the home flock, ducks used as breeders must be chosen carefully. Only birds displaying the following characteristics should be selected: robust health, strong legs, good body size, and freedom from deformities. If purebred ducks are raised, breed characteristics for typical size, shape, carriage, color, and markings should also be considered.

Characteristics to Avoid

The following deformities and weaknesses are highly inheritable and will occasionally show up in duck flocks. Birds with any of these faults *should not* be kept for reproduction.

Blindness

Certain strains of poultry, particularly some of those that are highly inbred, show a high incidence of clouded or white pupils. Some birds will exhibit this fault at hatching time, while in others it may not develop for several months or years.

Crossed Bill

When a duck has this defect, the upper mandible is usually bent to the side and not aligned with the lower mandible. If a duckling cuts the ridge of its bill on the eggshell at the time of hatching, the bill will often grow out crooked; this is not inheritable.

Curled Toes

The toes are bent so severely that the duck has difficulty walking. (Most ducks have slightly curved outer toes.)

Kinked Neck

This condition can be the result of an injury or from forcing tall ducks such as Indian Runners to remain in low boxes for extended periods of time (e.g., when they are shipped). Crested ducks exhibit necks with severe crooks just behind the head more than other breeds.

Roach Back

A deformed spinal column causes a humped and shortened back.

Scoop Bill

Scoop-bills have an unnaturally deep, concave depression along the top line of the bill.

Weak Legs

While this problem can usually be traced to nutritional deficiencies or obesity, weak legs can also be inherited. Genetic leg weakness is often indicated by birds hobbling about with one or both legs twisted slightly outward or birds whose legs give out quickly after walking short distances.

Wry Tail

Rather than pointing straight back as is normal, wry tails are constantly cocked to one side.

Catching and Holding Ducks

Ducks *must be handled with care* since their legs and wings are injured easily. When catching, avoid running them on rough ground or where they will trip over feed troughs and other obstacles. It is advantageous to walk waterfowl into a small catching pen or V-shaped corner rather than having a wild chase around a large yard.

Never grab ducks by their legs or wings. Rather, grasp them securely but gently by the neck or with one hand over each wing to subdue them, and then slide one hand under the breast and secure the legs. When the bird is lifted from the ground, its weight should be resting on your forearm with its head pointed back between your body and arm, and its wings pinned against your side. When the wings and feet are held securely, there is little possibility that either you or the duck will be injured. Small and medium-sized ducks can be picked up and held with a thumb over each wing and the hands encircling the body.

A safe and convenient method for catching ducks is to grasp the bird gently by the neck.

Small and medium-sized ducks can be caught and held with a thumb over each wing and the hands encircling the body.

Here are two methods for holding ducks that are comfortable and safe for both the bird and the holder.

Special precautions should be taken when handling Muscovies. They are surprisingly powerful and have long, sharp claws. When handling any type of bird, hold them away from your face to eliminate the possibility of any injury to your eyes. Wearing gloves to avoid cuts and scratches is also recommended.

Tiny Tina

Tiny Tina was my favorite boyhood duck. She would come running when I called her name and loved to follow me around the home place when the duckyard gate was left open. We would hike across the adjoining 80-acre field, and when it was time to return, she would fly home. Occasionally I would take her to the creek so she could take a dip and hunt water bugs.

Tiny was a dandy broody, and each spring for 6 years she laid a nest full of eggs and hatched them. The 7th year she laid a few abnormally small eggs and showed no interest in incubating them. Her voice started sounding hoarse and I was concerned that she was sick.

That summer when she molted, to my astonishment, her head became mottled with greenish-black and her sides turned gray. And there in the middle of her tail were several partially curled drake feathers! That fall Tiny Tina's name was officially changed to Tiny Tim.

Distinguishing the Sex of Mature Ducks

The gender of adult ducks is easily recognized by secondary sex characteristics such as voice, feather formation, plumage color, and body size. The voices of drakes resemble a hoarse "wongh", in contrast to the quacking of hens. Except during the molt, drakes display several curled tail feathers, and in colored varieties, they are brighter than hens.

Muscovies are practically mute, although the hens can quack weakly. Drakes lack the curled tail feathers of true ducks, and the two sexes are similar in color. However, the sexes can be readily identified by the drakes' larger size and the hens' smaller facial skin patch.

Clipping Wings

To keep flying ducks from wandering, it may be necessary to clip their wings. This is easily and painlessly accomplished by cutting the main flight feathers of one wing with tin shears or heavy-duty scissors. So that birds do not look unbalanced, the two outermost flights can be left intact. Once a year ducks molt their wing feathers and replace them with a new set. To keep them grounded, trim their wings annually.

You can ground flying ducks by clipping the main flight feathers of one wing, as shown here by the dotted line.

Reducing Manure Shoveling

Fortunately, the management of duck manure does not need to be a back-breaking chore. If you plan ahead, the birds will spread much of their droppings directly onto the land where it is needed.

In the fall of the year, after the garden crops have been harvested, a portion of, or the entire garden can be converted into the duck yard. A layer of bedding such as cornstalks, leaves, or straw should cover the ground to a depth of at least 3 to 4 inches. This covering not only keeps the ducks out of mud, but it also protects the dirt from being packed down and provides an excellent environment for earthworms and soil-enriching microorganisms.

In the spring, the partially decomposed bedding and manure can be worked into the soil. In gardens where this method has been used, lush crops of vegetables and fruits have flourished with minimal use of commercial fertilizers.

FIFTEEN

UNDERSTANDING FEEDS

To be healthy and productive, a duck must consume a diet that provides *all of the essential nutrients* in the *proper quantity* and in the *correct balance* with other nutrients. Because ducks are one of the hardiest and most adaptable of all domestic fowl, people often make the mistake of assuming that they can thrive on a poor diet. What a duck eats has a profound influence on its growth rate, appearance, productivity, and longevity. If fed an improper diet, even ducks out of world-class stock will not perform well or look like their parents.

Main Dietary Stages of Ducks

Wild ducks are continually adapting their diet to the special requirements of a particular season. Under domestic conditions, it is the responsibility of duck keepers to make sure their birds are consuming an appropriate diet for each of the six main stages of life.

1. Starting stage. For the first 2 weeks, ducklings require a higher level of protein and some vitamins than at any other stage. Starter feed must be in a size that the little birds can eat and easily digest. Recommended protein levels are 18 to 20 percent with a calcium:phosphorus ratio of approximately 1:1.

2. Growing stage. In weeks 3 to 8, ducklings are one of the fastest growing of domestic fowl. To sustain such rapid growth and minimize physical deformities, ducklings must consume a high-quality diet that meets all of their nutritional requirements but at the same time does not push them beyond a reasonable growth rate. For pets, breeding stock, and exhibition birds, a ration with 15- to 16- percent high-quality protein will encourage a moderate growth rate. For maximum growth rate a protein level of 18 percent is often used. The calcium:phosphorus ratio should be approximately 1:1. Do be aware, however, that the higher the protein percentage, the higher the likelihood of physical problems, such as wing and leg deformities and kidney and liver damage.

3. Developing stage. Depending on the breed and management, ducks have reached 70 to 90 percent of their adult weight by the time they are 9 weeks old. Between 9 and 20 weeks of age, they grow slowly, replace their juvenile plumage, and mature sexually. During this developing stage, a level of 13- to 14-percent high-quality protein is adequate with a calcium:phosphorus ratio of approximately 1:1.

4. Laying stage. Three weeks prior to and throughout the laying season, the requirements for protein, many vitamins, and some minerals (notably calcium) increase markedly. If the protein is high-quality and the essential amino acids carefully balanced, 15-percent crude protein is adequate. However, 16- to 17-percent protein is often used and 18- to 20-percent protein can result in larger eggs. (As temperatures increase, protein levels need to be raised if egg size is to be maintained.) Calcium requirements jump to 2.5 to 3 percent, while total phosphorus levels stay at about 0.65 percent.

5. Holding or maintenance stage. When fully mature ducks are not laying, their protein requirements drop to 12 to 14 percent with a calcium level of 0.6 to 1 percent and total phosphorus at 0.6 percent.

6. Molting stage. Each year, mature ducks typically go through one complete molt (eclipse) and one or more partial molts. Growing feathers require additional protein and other nutrients; therefore a well-balanced diet during the molt will encourage superior plumage. A level of 15- to 16-percent high-quality protein is adequate. The addition of animal protein (5- to 10-percent cat kibbles is a good source) and 10- to 20-percent oats during the molt can improve feather quality.

Types of Feeds and Supplements

When you go to a feed store, the many different kinds of feeds available can be confusing. To further complicate matters, different feed companies use different names for the same class of feed. Also, in some parts of North America, feed stores may not carry "duck feed." Here's a guide to help you understand common feeds and how to use them.

Duck or Waterfowl Starter/Grower

These feeds are formulated specifically for the dietary needs of baby ducks and geese. Because starter and grower feeds are similar, most feed companies make a combination starter/grower rather than two separate rations. These feeds typically have 18 to 20 percent protein.

If you are raising your ducks for meat and want the maximum growth rate, feed these as the sole ration. However, if you are raising your ducks for pets, breeding stock, or exhibition, and your duck starter/grower has more than 16 percent protein, the birds will normally live longer and have fewer leg and wing problems if their growth rate is slowed down by adding oats (either meal, rolled, whole, or pelleted) by 5 percent volume each week until the birds are receiving 3 parts starter/grower and 1 part oats.

Feed for Special Needs

Always make sure you provide a diet that is age- and life-stage-appropriate. For example, ducklings need a ration formulated specifically for baby waterfowl, whereas ducks that are laying require a diet that meets their special needs.

Chick Starter/Grower

Most feed stores sell chick starter/grower. These feeds are formulated for egg-type baby chickens, which have lower niacin requirements than ducks. If you use chick starter with ducklings, supplemental niacin should be supplied. Feed manufacturers can vary the amount of niacin they include in their feeds, but under most circumstances, satisfactory results are achieved if you add 100 mg of niacin per gallon of drinking water from 0 to 8 weeks of age for ducklings being fed chick starter/grower. Niacin is available in tablet or powder form at drugstores and health food stores. Keep in mind that excessively high levels of niacin can be toxic, so do not be tempted to put in extra for "good measure."

Broiler Starter and Meat Builder

Broiler starters are formulated for Cornish-cross meat chickens, which have niacin requirements similar to ducklings. Meat builder is manufactured for all types of fast-growing meat birds, including ducks. If they are nonmedicated or medicated with a drug that is not harmful to ducklings (such as amprolium or zinc bacitracin), these feeds can be used for ducklings. However, they are excessively high in protein (usually 20 percent or more) for pet, breeding, and exhibition ducks, so they should be supplemented with oats as outlined under Duck Starter.

Turkey or Gamebird Starter/Grower

Turkey and gamebird starter/growers have plenty of niacin for ducklings, but are excessively high in protein at 22 percent or higher. Because of their high protein content, they can cause a variety of physical problems (especially leg and wing deformities) and damage the kidneys and liver if fed as the sole ration to ducklings. Some people have satisfactory results by mixing 1 part chick starter, 1 part turkey or gamebird starter/grower, and ½ part uncooked oatmeal for a starting and growing ration for ducklings when duck starter/grower is not available.

Duck Developer

Developer rations usually contain about 14 percent protein and are designed to keep ducks between 9 weeks of age and sexual maturity healthy and in good feather condition without becoming obese. If a duck developer is not available from your feed store, you can use gamebird flight conditioner or mix your own by blending together 5 parts duck starter/grower (broiler starter/meat builder), 2 parts oats, 2½ parts wheat, and ½ part cat kibbles.

Common Causes of Increased Feed Consumption

1. Fast growth rate
2. Onset of egg production
3. Decreasing environmental temperatures
4. Wind chill
5. High-fiber feed with low caloric density
6. Heavy infestation of internal parasites
7. Increased exercise
8. Rodents or wild birds stealing feed
9. Decline in natural foods (grass, seeds, insects, invertebrates)

Gamebird Flight Conditioner

This feed is formulated to help gamebirds be in good feather and muscle condition for flying. With ducks it is useful as a developer, maintenance, and exhibition feed. A duck developer and duck maintenance ration can be made by mixing 4 parts gamebird flight conditioner with 1 part oats.

Layer Rations

Layer rations are formulated for birds that are producing eating eggs. (If eggs are going to be hatched, a breeder ration should be used.)

To lay well and not deplete the body of minerals (especially calcium) and vitamins, ducks need to eat a laying feed 2 to 3 weeks prior to and throughout the laying season. Most good chicken laying rations work satisfactorily for ducks. While conducting feed trials with laying ducks, we had a Khaki Campbell lay 357 eggs her pullet year on a 16-percent-protein chicken layer manufactured by a national feed company.

Warning: Because of the high calcium levels (2.5–3.5% Ca) of laying feeds, they should *never* be used for growing birds. More than 1.5 percent calcium in the diet of nonlaying birds can cause permanent damage to organs and the skeleton, or even death. When possible (and this can be difficult), drakes should not consume laying feed continuously during the laying season. At the conclusion of the laying season, ducks should be switched to a maintenance feed containing 0.60 to 1 percent calcium.

Breeder Rations

These feeds are formulated for birds that are producing eggs for hatching. When compared with layer feeds, breeder rations normally have higher levels of some nutrients such as vitamin E, riboflavin, pantothenic acid, pyridoxine, biotin, folacin, and the minerals copper, iron, manganese, and zinc. In general, calcium levels are about 0.5 percent lower in breeder rations when compared with layer rations. Breeder rations are designed to produce high fertility, strong embryos, and good hatchability. Gamebird breeder rations often work well with ducks — especially with strains that have a history of poor fertility and hatchability.

When breeder rations are not available, the following mix will usually give good results: 8½ parts chicken layer, 1 part cat kibbles, and ½ part

Animax or Calf Manna (or similar product). Like layer feeds, breeder rations should be fed 2 to 3 weeks prior to and throughout the laying season, and discontinued promptly when egg production ceases.

Complete Waterfowl, All-in-One Waterfowl

To keep things simple, some feed companies produce one waterfowl feed that they market as appropriate for all ages. Basically these rations are starter/grower feeds that must be supplemented with calcium for laying birds — crushed oyster shells *and* ground limestone should be fed free-choice during the laying season when these feeds are used. Because these feeds are essentially starter/grower rations, they can usually be supplemented with grains during some stages of a duck's life. For example, grains can be added at the following rates: 10 percent grains added from 3 to 8 weeks of age and 20 to 25 percent from 9 weeks until point of lay and for adult ducks during the nonbreeding/laying season.

Maintenance Rations

Feed manufacturers seldom sell poultry/waterfowl maintenance rations. However, gamebird maintenance feeds are available in some localities and normally work satisfactorily for ducks. Also, gamebird flight conditioner is used by some duck keepers with good results as a maintenance ration.

You can also make a maintenance ration by mixing 25 to 50 percent cereal grains with starter/grower feeds. How much grain should be added depends on the particular starter/grower feed used and on the environmental temperature. As a general rule of thumb, when temperatures are below freezing, 50 percent grains can be added, and when temperatures are above 80°F, no more than 25 percent grains should be added.

Concentrate

Manufactured concentrates such as Calf Manna (Manna Pro) and Animax (Purina) are easily digested and have a high concentration of protein, vitamins, and minerals. These products can be useful during times when supplements are needed (the breeding season, for example) or when feed consumption is down due to old age, stress, injury, or high environmental temperatures. They are normally used at 5 to 10 percent of the ration.

Due to their high concentration of nutrients, care must be taken to not feed excessive amounts, which could lead to toxic overdose.

Cat Kibbles

High-quality cat kibbles are especially useful as a supplement during the breeding season, helping ducks to grow glossy feathers for exhibition and maintaining body heat during cold weather. Like any unfamiliar feed, ducks often will not eat them at first. Due to the relatively high fat content of kibbles, they are usually fed at no more than 10 to 15 percent of a duck ration for a prolonged period of time.

Vitamin and Mineral Supplements

Vitamin and mineral supplements are available in powder, liquid, capsule, and tablet form. Vitamin and electrolyte preparations that are formulated for poultry and gamebirds can be useful for getting newly hatched ducklings off to a good start and boosting the immune system of birds that are ill, shipped, or stressed by old age or severe weather. Some vitamins and minerals are toxic when ingested in excessive quantities, so these supplements must be used with care.

Antibiotics

The proper use of antibiotics treats and reduces the occurrence of some diseases. Antibiotics work by killing bacteria in the body; unfortunately, they destroy both "good" and "bad" microorganisms. While they are useful for treating some diseases, their overuse has profound long-term negative consequences, including decreased efficacy of the drug, contaminated meat and eggs, and decreased natural resistance in a flock of birds. Always use antibiotics with care and follow withdrawal recommendations carefully.

Probiotics

Whereas antibiotics destroy microorganisms indiscriminately, probiotics are actually beneficial bacteria that boost the body's natural defenses against some diseases. In healthy animals, the "good" microorganisms are able to keep the "bad" ones from multiplying sufficiently to endanger the health of the host.

Research indicates that supplementing the diet of animals with probi-otics results in healthier, more productive birds. Probiotics are especially helpful for getting newly hatched ducklings off to a good start and as a sup-plement to the diet of ducks that are old or otherwise stressed. We add V.M.P. Poultry Pac (Vitamins, Minerals & Probiotics), manufactured by RXV Products, to the drinking water of our ducklings for the first 5 to 10 days after hatching.

Granite Grit

To aid their gizzards in grinding food, ducks need to ingest insoluble grit such as crushed granite or coarse river sand. Feed stores sell granite grit in various sizes: The chick-sized version should be used for young ducklings, and the hen-sized for ducks six weeks and older.

Crushed Oyster Shells

Crushed oyster shells are commonly used to supplement the diet of birds with calcium. Because oyster shells are hard and dissolve relatively slowly, they also function as a grit. An advantage of oyster shells is that they provide a steady supply of calcium to the bloodstream 24 hours a day. This metering effect is especially important to laying ducks during warm and hot weather.

Growing ducklings and nonlaying adults should be fed oyster shells *only* if it is known that their diet is deficient in calcium. Oyster shells can give ducklings that are already receiving adequate calcium a harmful calcium overdose. Oyster shells are available in both chick and adult sizes.

Grains

Grains make up the majority of most duck rations. However, by them-selves they are not a nutritionally complete diet. Many nutritional deficien-cies result from people assuming that ducks can thrive solely on a diet of grain. Worldwide, many kinds of grains are fed to ducks. Here, we will briefly look at the ones most commonly used in North America.

Corn

Corn is fed more than any other grain. It is high in energy, which makes it valuable as a cold-weather feed, but also tends to make ducks fat. Ducks are fond of corn and will overeat it. Corn is susceptible to molds that produce aflatoxins, which can be harmful or fatal to ducks. Never feed corn (or any grain) that looks or smells moldy.

Wheat

Wheat is higher in protein and lower in energy than corn. When hard wheat is fed prior to being sufficiently cured (45 to 90 days), some duck raisers have reported that their birds developed severe diarrhea. On the other hand, we feed tons of soft white wheat straight out of the field every summer and fall with no trouble. Ducks are fond of wheat and tend not to get as fat on it as corn.

Oats

Oats are lower in energy and have more than four times the amount of fiber that corn and wheat have. Most ducks initially are not fond of whole or rolled oats, and may take some time to learn to eat them. In general, ducks that have 5 to 25 percent oats in their diet grow slower, have fewer leg and wing deformities, have better-quality feathers, are less likely to be fat or eat feathers, produce more offspring, and live longer. Due to the high fiber content of oats, it normally is not advisable to include them at more than 5 to 25 percent of the ration.

Soybeans

Most poultry feeds use heat-processed soybean meal as a protein supplement. Raw soybeans are unsatisfactory because they are unpalatable and cause retarded growth.

Texture of Feed

Feeds are available in a variety of textures. Each has advantages and disadvantages in how they are prepared and consumed.

Mash

In mash feeds, ingredients are ground up into a coarse, medium, or fine flour consistency. Generally, grinding makes feeds easier to digest and promotes even mixing of the ingredients. Mash feeds are dry and powdery, causing birds to eat them more slowly (which can be either an advantage or disadvantage), and in the case of ducks, increases spillage and waste.

When fed to ducks, mash can be moistened (not enough to be soupy) with water or milk, but if this method is used, only mix up what will be cleaned up in a few hours. Moistened feeds spoil rapidly and, if not carefully monitored, can cause serious problems. Because of the small particle size of mash, ducks of any age can eat it.

Pellets

Many feed companies run their mash through a pellet mill to make a firm pellet in order to reduce dustiness and waste. Pelleting is an additional expense in manufacturing feed, but the reduced loss of feed through spillage usually more than makes up for any price increase. Most pellets are too large for ducklings (especially smaller breeds) to consume for at least the first week or two. Because pellets are easy for ducks to pick up, they tend to eat more feed in this form than finer-textured feeds.

Crumbles

Feed that has been pelletized can be crushed to form crumbles. The main advantage with crumbles is that birds of all ages can eat them, and they also slow feed consumption slightly when compared with pellets. There normally is more waste with crumbles than with pellets.

Whole Grain

Once they are familiar with them, ducks will consume many types of whole grains. Ducklings have difficulty eating all but the smallest grains. Due to a protective hardened outer covering, whole grains are more resistant to nutrient loss and spoilage than in any other form. Whole grains are harder to digest (primarily a concern with very young or old birds or in situations where no insoluble grit is provided) and when used in mixed rations, ducks tend to pick out their favorites and leave behind the rest.

Cracked, Rolled, Crimped, Flaked Grains

When the protective outer coating of grain is broken, the nutrients are often easier to digest. However, the grain also becomes more susceptible to nutrient loss and spoilage. Therefore, freshness is critical.

SIXTEEN

BUTCHERING

One of the main reasons ducks are raised is for their excellent meat. While they can be taken to a custom dressing plant to be butchered, this service is often expensive and robs you of the satisfaction of preparing your own food.

Cleanliness throughout the butchering process is essential to curb contamination and spoilage of meat. All cutting utensils should be sharp — dull knives are a waste of time as well as unsafe.

When to Butcher

One of the most time-consuming parts of the butchering process is the removal of the feathers. To make this job as simple as possible, butcher ducks when they are in full feather. If a duck is slaughtered when it is covered with pinfeathers, a picking job that normally takes 3 to 5 minutes can develop into a frustrating, feather-pulling marathon.

Depending on the breed and management, ducklings are normally in full feather for only 5 to 10 days sometime between the age of 7 to 10 weeks of age, except Muscovies, which require 14 to 16 weeks to feather out. Shortly after achieving full feather, young ducks go into a molt and begin replacing their juvenile garb with adult plumage. If ducklings are not dressed before this molt commences, butchering is best delayed for 6 to 10 weeks when their adult plumage will have been acquired.

Preparations for Butchering

Ducks should be taken off feed 8 to 10 hours prior to killing, or the night before if they are going to be dressed early the following morning. To avoid excessive shrinkage, drinking water should be left in front of the birds until they are slaughtered.

Killing

For most of us, the least enjoyable task in raising ducks is killing them on butchering day. But for those of us who raise our own meat, it's a necessary chore.

There are several ways of dispatching ducks. The simplest and most impersonal method is the ax and chopping block. It is advantageous to have a device (such as two large nails driven into the block to form a V) on the chopping block to hold the bird's head securely in place. A sharp cutting edge on the ax is a must. As soon as the head is removed, the bird should be hung by its legs to promote thorough bleeding and to prevent it from becoming bruised or soiled from thrashing about on the ground.

A second method of killing is to suspend the live duck head-down with leg shackles or in a killing cone. Grasp the bill firmly with one hand; with the other stun the bird with a stout stick. With a sharp knife, cut the throat about

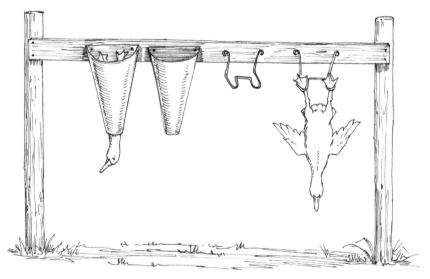

Killing cones and leg shackles allow the ducks to hang head-downward during killing.

an inch below the base of the bill, on the left side, severing the jugular vein. The head provides a convenient handle if the duck is going to be scalded.

To avoid discolored meat in the dressed product, it is vital that ducks are bled *thoroughly* before they are processed further.

Picking

The sooner a duck is picked after it is bled, the easier the feathers will come out.

Dry Picking

The best-quality feathers for filler material and the most attractive carcasses are obtained when ducks are dry picked. Novices often find this method unbearably slow. Veterans, on the other hand, can dry pick a duck clean in 3 minutes or less. One secret for success is to extract feathers in the same direction they grow. Pulling feathers against the grain invariably results in torn skin. For a better grip, lightly dust the duck with resin, or periodically moisten your hands with water. The large plumes of the wings and tail need to be plucked out one or two at a time.

Because feathers and down float in every direction when waterfowl are dry picked, choose a setting that is free from drafts. Picking into a large plastic garbage bag and covering the floor around the plucking area with a tarp or an old sheet can help keep feathers under control.

Scald Picking

The most common method for defeathering ducks is to scald them prior to picking. The only equipment needed is a large container in which to submerge an entire duck, and a forked branch (or a 1-inch × 2-inch × 24-inch board with a large nail driven in at an angle near one end) which is used to hold the bird under water.

To scald, hold the duck by its feet and wedge the neck into the fork of the stick. Then dip the bird up and down in water that is 125 to 145°F, making certain that the water penetrates through to the skin. To improve the wetting ability of the scalding water, a small amount of liquid dishwashing detergent can be added.

The scalding time for ducks varies from 1 to 3 minutes, depending on the age of the bird and the temperature of the water. Mature ducks require longer and hotter scalds than ducklings. If you find that the feathers are still difficult to remove after the initial scald, the bird can be redipped. However, overscalding causes the skin to tear easily and discolors the carcass with dark blotches.

Ducks should be picked immediately after scalding, starting with the wing and tail feathers. Because the feathers are going to be hot, have a bucket of cold water nearby to dip your hands into occasionally.

If your first attempts at scald picking poultry do not produce carcasses that are as attractive as those that are processed commercially, don't get discouraged. Your results should improve once you've gained a little experience. Feathers from scalded ducks are of good quality when handled correctly (see Care of Feathers, page 233).

Wax Picking

A popular variation of the scalding method is to dip rough-picked ducks in hot wax. Even when birds with a moderate number of pinfeathers are butchered, this procedure can produce clean carcasses in a short time. Paraffin or a mixture of 1 part beeswax to 1 part paraffin can be used. There are also products such as Dux-Wax that are made specifically for this purpose and are available from some poultry supply distributors. (See appendix G, page 299.)

When wax is used, ducks are scalded without detergent in the water, and are rough-picked by removing the large tail and wing feathers and 50 to 90 percent of the body plumage. Prior to scalding, place a container of solidified wax in a large receptacle of hot water where it is melted and heated to a temperature of 150 to 165°F. *Extreme care* must be taken when working with hot wax to avoid burns and fires.

Rough-picked ducks are partially dried and then dipped into the hot wax several times. Spray with cold water or wait long enough between each dunking to allow sufficient congealing that will build up a good layer of wax. If only a few birds are being dressed, you may find it simpler to melt a small container of wax to be poured over the carcasses.

Submerging the waxed duck in cold water causes the wax to harden and grip the feathers. The wax and feathers are then stripped off together, resulting in a finished product that is clean and attractive. Birds that are extra pinny can be rewaxed.

Used wax can be recycled by melting it and skimming off the feathers and scum, and boiling out any existing water.

Singeing

After ducks are picked, long, hair-like filament feathers usually remain. The simplest way to remove these filoplumes is to pass the carcasses quickly over a flame, being careful not to burn the skin. A jar lid with a thin layer of rubbing alcohol in the bottom gives the best flame I know of for singeing. Alcohol burns tall, cleanly, and odor-free. Newspapers (do not use colored sheets) loosely rolled into a hand torch, gas burners, and candles can also be used.

Skinning

Ducks can be skinned rather than picked. Some advantages of this technique are that ducks with pinfeathers can be dressed as easily as those in full feather and some people find skinning less time-consuming than picking. Because skin is composed largely of fat, skinning significantly reduces the fat content of the dressed duck.

The major drawbacks of this method are that skinned carcasses lose much of their eye appeal when roasted whole, special precautions must be taken in cooking the meat to prevent dryness, some flavor is lost, and a higher percentage of the bird is wasted.

To prepare a duck for skinning, remove its head, feet, and the last joint of each wing. With the bird resting on its back, slip the blade of a small, sharp knife under the skin of the neck, and slit the skin the length of the body, cutting around both sides of the vent. The final step is to peel the skin off, which requires a good deal of pulling. Over stubborn areas, a knife is needed to trim the skin loose.

Eviscerating

Ducks can be drawn immediately after they have been defeathered, or they can be chilled in ice water for several hours or hung in a cool (33 to 36°F) location to ripen for 6 to 24 hours. Chilling the carcasses first has the advantage of making the cleaning procedure less messy, while aging before eviscerating produces stronger-flavored meat, which is preferred by some people.

With the bird resting on a clean, smooth surface (we use a decommissioned cookie sheet), remove the feet, neck, and oil gland, making certain that *both* yellowish lobes of the oil gland are cut out. Then make a shallow 3-inch-long horizontal incision between the end of the breastbone and the vent, being careful not to puncture the intestines that lie just under the skin.

Through the incision insert your hand into the body cavity and gently loosen the organs from the inside walls of the body, and pull them out. Cut around the vent to disconnect the intestines from the carcass. The gizzard, heart, and liver can be cut free and set aside before the unwanted innards are discarded.

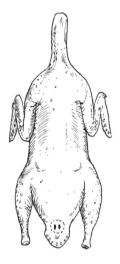

Cut off both lobes of the oil gland located on top of the tail.

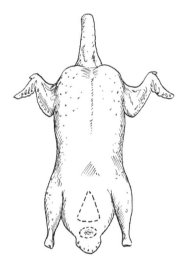

Make an incision between the end of the breast bone and the vent to remove innards.

The esophagus (ducks do not have true crops) and windpipe are well anchored in the neck and require a vigorous pull to remove them. The pink, spongy lungs are located against the back among the ribs and can be scraped out with the fingers if you desire.

To clean gizzards, cut around their outside edge and then pull the two halves apart. The inner bag with its contents of feed and gravel can then be peeled away and discarded. The final step is to rinse the muscular organ with water.

The gallbladder, a small green sac, is tightly anchored to the liver and should be removed *intact* since it is filled with bitter bile. A portion of the liver must be cut off along with the gallbladder so bile does not spill onto edible meat.

Unwanted feathers and body parts make excellent fertilizer and should be buried near a tree or in the garden. *Do not feed raw innards to cats and dogs.* If you give uncooked entrails to your pets, they can develop a taste for poultry and may kill birds to satisfy their cravings. Also, simply throwing the innards across the fence will likely draw predators from near and far.

Cooling the Meat

After all the organs have been removed, thoroughly wash the carcass and chill it to a temperature of 34 to 40°F as soon as possible. If meat is cooled slowly, bacteria may grow, causing spoilage and unpleasant flavor. Poultry can be chilled in ice water or air-cooled by hanging the carcasses in a refrigerator or in a room with a temperature of 30 to 40°F.

Packaging and Storing Meat

After the body heat has dissipated from the carcasses, they should be sealed in airtight containers. If the meat was chilled in water, it should be allowed to drain for 10 to 20 minutes before it is packaged. To retain the highest quality in meat that is going to be frozen, suck the air from plastic bags with a straw or vacuum cleaner before sealing. To produce the tenderest meat, poultry must be aged 12 to 36 hours at 33 to 40°F before it is eaten or frozen.

Care of Feathers

Duck feathers are a valuable by-product of butchering. If you plan to save the feathers, keep the down and small body feathers separate from the large stiff plumes of the wings, tail, and body as the slaughtered birds are being picked.

When ducks are scalded prior to picking, the feathers need to be washed with a gentle detergent, rinsed thoroughly in warm water, and spread out several inches thick on a clean, dry surface or loosely placed in cloth sacks and hung in a warm room. Stir the wet feathers daily to fluff them and ensure rapid drying. Once they are well dried, feathers can be bagged and stored in a clean, dry location. (See appendix E, page 296, for instructions on how to use feathers.)

Butchering Checklist

❑ Remove the ducks' feed 8 to 10 hours before they are killed.

❑ Sharpen all cutting tools.

❑ When catching ducks prior to butchering, keep them calm and handle them carefully to avoid discolored meat due to bruises.

❑ Hang slaughtered ducks in killing cones or by their feet to avoid bruising or soiling the meat and to ensure thorough bleeding.

❑ Remove feathers as soon as possible after birds are bled.

❑ Singe off filament feathers.

❑ Trim out the oil gland from the base of the tail.

❑ Cut off feet and shanks at hock joints; remove neck.

❑ Make a horizontal incision between the vent and the end of the breastbone, and gently lift out innards.

❑ Set aside the heart, gizzard, and liver before disposing of unwanted innards.

❑ Pull out the windpipe and esophagus from the neck/chest area.

❑ Extract the lungs from between the ribs.

❑ Rinse the carcass thoroughly with clean, cold water.

❑ Clean the gizzard and carefully remove the gallbladder from the liver.

❑ Chill dressed duck to an internal temperature of 40°F.

❑ Bury unwanted body parts, entrails, and feathers in the garden or near a tree, sufficiently deep so that scavengers will not dig them up.

❑ Age the meat for 12 to 36 hours at 33 to 40°F prior to cooking or freezing.

❑ Package and freeze the meat, or enjoy a festive banquet of roast duck.

Seventeen

Health and Physical Problems

Compared with chickens, ducks have greater resistance to many diseases and parasites. Providing ducks with a proper diet, reasonably clean drinking water, adequate shelter, and sufficiently clean living quarters with ample space minimizes health problems. In my 40 years of raising thousands of ducks and every recognized breed, I have never had to treat a home-raised broadbill for coccidiosis or internal parasites, or had to use a vaccine for a communicable disease.

The most common causes of health problems in ducks are improper diet, ingestion of toxic substances, overcrowding, filthy pens, and injuries. If you are unable to diagnose a malady, promptly seek advice from an experienced waterfowl breeder, veterinarian, or animal diagnostic laboratory. Procrastination may prove fatal and can result in an epidemic. Many infectious diseases cannot be accurately diagnosed without proper lab tests. Check with your local veterinarian for the nearest diagnostic lab.

Sick or injured ducks should be penned in a dry, warm, clean enclosure (if a contagious disease is suspected, the sick pen should be isolated, well away from other birds), and provided a balanced diet and clean drinking water. Dead birds should always be removed as quickly as possible to avoid attracting predators and, for communicable diseases, to curb the likelihood of infecting healthy birds. Burn carcasses in an incinerator or bury them deep enough that they will not be dug up by scavengers. Leaving deceased birds laying around or tossing them over the fence into the bushes is an invitation for trouble.

Categories of Health and Physical Problems

Based on their origin, health and physical problems can be grouped into five main categories: nutrition and toxic substances, infectious organisms, injuries, parasites, and genetic defects.

Nutrition and Toxic Substances

The leading causes of health and physical problems in ducks are improper diet and/or ingestion (and rarely, inhalation) of toxins. Therefore, if ducks in your flock are experiencing health problems, but display no obvious signs of injury or infectious disease, the first step is to determine if diet is the culprit. (I am using the term *diet* to include anything the birds ingest, including contaminated soil and water and poisonous plants and animals.)

To determine if health problems are related to something the ducks are ingesting, you can: 1) pen birds in a clean pen where you can be sure they are not consuming poisonous plants, invertebrates, baits or sprays, leaded paint, or contaminated soil or water; and 2) change their feed to a well-known brand that is balanced for all nutrients. (In my experience, Purina typically has high quality-control standards and balanced formulas.)

The most common causes of health problems related to diet include: 1) nutritional deficiencies; 2) nutritional overdoses (where vitamins and minerals are toxic in excessive amounts); 3) nutritional imbalances (the ratio of some nutrients to each other is as important as the quantity — phosphorus and calcium for example); 4) medication toxicity; 5) molds; 6) spoiled feed; 7) toxic plants or animals; 8) pesticides (which often are harmful to ducks *even if* the manufacturer claims otherwise); 9) baits; 10) leaded paints or solvents; 11) contaminated water or soil; and 12) rotting plant or animal material.

A common mistake is to presume that a health problem cannot be diet-related because only part of the flock is showing symptoms. *Remember, every bird digests and assimilates nutrients and toxins differently* and it is normal for diet-related symptoms to vary widely in a flock of birds, from individuals that show no symptoms to others that experience severe symptoms or death. Furthermore, some breeds, varieties, and even strains within a variety have, as a group, differing dietary needs and sensitivities. A diet that is adequate for one duck may have serious consequences for another.

Here are some examples of diet-induced health problems in ducks.

Salt Deficiency

A few years ago, a number of waterfowl raisers along the East Coast reported that their young birds showed symptoms including listlessness, irregular growth, poor feather quality, and, in some cases, increased mortality. It was then observed that everyone whose birds were affected used the same brand of grower feed. A feed analysis showed that several vitamins were marginally low, and the salt content was well below the minimum requirement. The feed company later acknowledged that salt had been inadvertently omitted.

Excessive Calcium

One year, about 10 percent of our 5- to 8-week-old ducks began looking listless, and small spots of blood were discovered in their droppings. I immediately suspected coccidiosis and rushed several birds to the veterinary diagnostic lab at Oregon State University. The lab report came back negative for coccidiosis or for any infectious disease. I called the mill that custom mixes our feed and they assured me that the correct formula had been used.

After a second negative lab report, I switched the birds to a national brand of grower feed and within a few days, the symptoms gradually disappeared and there was no more mortality. After our feed company did further checking, they discovered they had accidentally doubled the calcium level from 1 to 2 percent in the last batch of our grower pellets. Therefore, the growing birds were suffering from a phosphorus-calcium imbalance, as well as excessive calcium.

Mycotoxin-Contaminated Feed

Mycotoxins are produced by molds. When excessively high levels are in their feeds, some ducks in a flock will not thrive and mortality rates will increase.

Vitamin Deficiency

A number of years ago, a company sold a feed that it advertised as being adequate for all species and ages of poultry. When this ration was fed to ducklings, some of them grew slowly and/or developed bowed legs. When an independent lab analyzed the feed, it found that for ducklings, it contained borderline deficiencies for several vitamins, including niacin. Just because a ration is recommended by a company or dealer does not guarantee that it is adequate for ducks.

Infectious Organisms

When ducks are raised in relatively small flocks with plenty of room and a calm environment, and consume a proper diet that supplies all essential nutrients in adequate quantities, they are impressively resistant to infectious diseases. Ducks that are stressed by crowding, filthy living conditions, or inadequate diet are more prone to succumb to infectious diseases.

Large commercial duck operations are more likely to encounter infectious diseases due to the high concentrations of birds. It sometimes becomes necessary for commercial operators to routinely medicate and/or vaccinate for highly infectious diseases.

Infectious diseases can be spread via wild birds and animals, infected ducks from other flocks, hatching eggs, clothes and footwear, and air or water. Some infectious diseases spread overnight while others may take weeks to move through a flock. Quick diagnosis and treatment are important for reducing severity and mortality.

Medications and antibiotics should always be used with care and only when necessary to minimize accidental toxicity or the development of resistant strains of disease organisms. When antibiotics are used it is important to complete the recommended treatment, even if the birds appear healed prior to completion. Some medications and antibiotics are toxic to some strains of ducks and not others; it is safest to administer medication to a few birds and look for negative reactions before treating an entire flock.

To find out if there are infectious diseases that are common in your area, consult with local duck raisers and veterinarians, extension agents, and regional diagnostic laboratories.

Injuries

Injuries are kept to a minimum when ducks are protected from predators, are kept in pens free of sharp objects and obstructions to trip over, are provided a calm environment, and handled properly and at a minimum. Due to their anatomical structure, the hips, legs, and feet of ducks are the most susceptible to injury. People often ask if they should "put down" a bird that has suffered major injuries. In my experience, birds appreciate the opportunity to recover. Ducks have an amazing ability to heal, even from injuries that appear to be catastrophic.

Parasites

Under normal circumstances, ducks have good to excellent resistance to most internal and external parasites. However, ducks that consume an inadequate diet or are crowded or forced to live in a filthy environment are more susceptible to parasites.

Genetic Defects

No matter how carefully breeding stock is selected and managed, some offspring will be produced with genetic physical defects. By definition, purebred animals are at least mildly inbred (this homogamy is the reason individuals of the same breed share similar characteristics). Therefore, they are inclined to have more genetic defects than hybrids. Generally, the more intensely a strain is selected for specific characteristics, the more frequently genetic defects appear. When raising young stock, do not be surprised if you need to cull out some defective specimens.

Bio-Security

In this day and age of increased travel, microorganisms commonly hitchhike on human or wild bird and animal hosts from places far and near. Therefore, it becomes increasingly important to minimize the spread of disease organisms. As the cliché goes, "An ounce of prevention is worth a pound of cure." Some commercial operations have elaborate bio-security measurements, but even small-flock owners can use the following strategies to greatly reduce the spreading of disease organisms and the risk of dangerous infections among their ducks.

Within a farm, footwear should be changed or washed prior to entering the young-bird area since ducklings are more susceptible to disease organisms than are older stock. Dead birds should be incinerated or deeply buried. Wild birds and rodents are potential disease and parasite carriers, so be sure to keep them out of buildings and feeders.

Because diseases are commonly spread on clothing and footwear, visitors to farms should not be allowed near or in the bird pens (unless they are provided clean overclothes and footwear). If you are visiting a bird farm, never assume you will be allowed into the pens.

Quarantining Birds

Birds that look healthy can be carriers of disease and parasites. Small-flock owners should treat new adult birds for internal and external parasites and quarantine them for 3 to 4 weeks, observing them for signs of disease prior to introducing them into the flock.

For further security, commercial farms and small-flock owners should keep new arrivals isolated for a minimum of 90 days since some diseases have a lengthy incubation period. For maximum security, birds can be quarantined for a full year and their eggs hatched in separate facilities. This provides protection against diseases transmitted via eggs. After caring for quarantined birds, change or wash your footwear before returning to the main facility.

First-Aid Kit for Ducks

♦ Antibiotic ointment/bacitracin for minor cuts and abrasions
♦ A medicated ophthalmic ointment for eye injuries and infections
♦ A broad-spectrum antibiotic (such as amoxicillin, tetracycline, or penicillin) to aid in the healing of infections
♦ A poultry or bird vitamin mix for adding to drinking water during times of stress
♦ A pyrethrin-based insecticide for treating external parasites and spraying around wounds to prevent maggots

Working with Your Veterinarian

A useful tool in maintaining a healthy duck flock is a good working relationship with a veterinarian. They can prescribe medications, help diagnose problems, do emergency surgery and injury repair, and provide information regarding infectious organisms. There is a growing number of avian specialists in the veterinary field. To find one near you, contact the Association of Avian Veterinarians by E-mail at aavctrofc@aol.com.

Diseases, Physical Disorders, and Parasites

Of the following ailments, those marked with an asterisk (*) are the most commonly seen disorders in small duck flocks.

*Aspergillosis**

Aspergillosis is a disease affecting the lungs, primarily of young ducklings. It is indicated by gasping or labored breathing, poor appetite, and general weakness, and occasionally is accompanied by sticky eyes. The lungs of affected birds often contain small yellowish nodules about the size of a BB shot. Ducklings are most susceptible the first few days after hatching.

Causes
Commonly known as *brooder pneumonia*, it is the result of *Aspergillus fumigatus* mold being inhaled into the lungs when ducklings are hatched in contaminated incubators or from being brooded on moldy bedding or fed moldy feed.

Treatment
There is no known cure, but to prevent its spread, infected birds should be removed, the brooding area and equipment disinfected, and the feed and bedding checked for musty odor or signs of mold.

Prevention
Hatch eggs only in thoroughly disinfected incubators and use mold-free bedding and feed.

Black Flies and Leucocytozoon Disease

This health problem is most common in Canada and subarctic zones, and primarily affects young, unfeathered birds. Sudden death is the most common symptom in ducks. Lab tests (blood smears) are required for positive identification.

Cause
Black flies (*Simuliidae*) are bloodsuckers and transmit Leucocytozoon disease.

Treatment and Prevention
Prevention is the only effective treatment I am aware of in ducks. People in areas where black flies are problematic have found that by having ducklings well grown by the time black flies emerge in the spring, most losses can apparently be avoided.

*Botulism**

Botulism, commonly known as limberneck, is usually fatal and can affect birds of any age. A few hours after eating poisoned food, birds may lose control of their leg, wing, and neck muscles. In some cases, body feathers loosen and are easily extracted. Ducks that are swimming when paralysis of the neck develops often drown. Dying birds may slip into a coma several hours before expiring. Botulism normally kills in 3 to 24 hours, although in mild cases birds may recover in several days.

Causes

This deadly food poisoning is caused by a toxin produced by *Clostridium botulinum* bacteria, which are commonly found in soil, spoiled food, and decaying animal and plant matter. Botulism strikes most frequently in dry weather when levels in ponds and lakes drop, leaving decaying plants and animals exposed for ducks to eat. Maggots that feed on decaying carcasses often carry the botulism toxin. Ducks can also contract this toxicant from spoiled feed or canned food from the pantry.

Treatment

All ducks suspected of having eaten poisoned food should be confined to a clean, shady yard or building and immediately provided fresh drinking water with a laxative added — either 1 pint of molasses or 1 pound of Epsom salts per 5 gallons water. Birds that cannot drink on their own should be treated individually. The addition of 1 part potassium to 3,000 parts drinking water or individual doses of 1 teaspoon castor oil have also been recommended as treatments. In birds that are particularly valuable, flush out the contents of the esophagus with warm water by using a funnel and rubber tube inserted into the mouth and several inches down the esophagus. To avoid further problems, every effort must be made to locate the source of botulism. A vaccine has been developed, but it is rather expensive and often difficult to obtain on short notice.

Prevention

Bury or burn carcasses of dead animals and clean up rotting vegetation. Do not let your ducks feed in stagnant bodies of water or give the birds spoiled canned goods or feed.

Broken Bones

The bones of birds have a wonderful ability to mend themselves. However, to prevent a duck from being permanently disfigured or crippled, it is often helpful to set and immobilize a wing or leg that is fractured.

Setting
Broken bones should be treated promptly, within 24 hours after the accident. A bone is set by gently pulling apart, and if necessary, slightly twisting the two halves until they mesh properly.

Splints
Broken bones should be held in alignment with splints (popsicle or tongue sticks often work well). A rigid support should be positioned on either side as far above and below the fracture as possible, and held securely in place with strong tape. The patient should be checked frequently to ensure that the brace is staying in place and blood circulation is not being restricted. Splints can normally be removed in 14 to 28 days.

*Broodiness**

Broodiness normally is not a sign of illness, although duck-raising novices are often alarmed by the behavior of broody females. The duck stays on her nest for long periods, quacks loudly when disturbed or when off the nest to eat, drink, and bathe, becomes protective of the nest, retracts neck and head down onto her shoulders, and defecates a large amount of foul-smelling excrement soon after leaving the nest or while on it if startled.

Cause
Broodiness is caused by physiological changes in female birds that cause an increase in body temperature and give her a strong urge to set on and protect a nest and/or brood babies.

Setting females are susceptible to external parasites such as mites, lice, and stinging ants, and she and her nest should be treated with an appropriate insecticide if needed. Females allowed to set on a nest well past the normal incubation period are susceptible to nutritional deficiencies and may waste away and die if not removed to a well-lit pen with no nests and provided a balanced diet and bathing water.

Choking*

Ducks will occasionally get feed caught in their throats. Normally, after a vigorous shaking of the head, the passageway is cleared and breathing returns to normal. However, sometimes a bird is unable to clear its throat and will suffocate if not promptly aided. When a duck obviously needs your assistance, pull its head forward until it is in a straight line with the neck, open the bill by squeezing with the thumb and index finger on both corners of the bill, and push your finger, a piece of ½-inch rubber tubing, or the eraser end of a new pencil down the bird's throat until the obstruction is dislodged.

Chronic Respiratory Disease (CRD, Mycoplasmosis)

Chronic Respiratory Disease is an infection of the respiratory system and is most common in times of stress. (See also Sinus Infection.) Symptoms include coughing, sneezing, thick mucous discharge from the nostrils, ruffled feathers, and irregular or stunted growth. Instead of all birds being infected at once, CRD tends to move through a flock rather methodically over the course of several days to weeks.

Cause
CRD is caused by the highly contagious microorganism *Mycoplasma gallisepticum* (MG). CRD outbreaks are most common after birds have undergone stress due to being shipped, moving to new pens, sudden feed changes, onset of egg production, drastic weather changes, diet, and such.

Treatment
At the first sign of diagnostic symptoms, administering an appropriate antibiotic can provide relief. Most strains of MG are sensitive to various antibiotics such as erythromycin (sold as Gallimycin) and tetracyclines (Aureomycin and Terramycin). Tylosin (Tylan) is often the most effective, although it can be hard to locate (ask your veterinarian).

Prevention
The best ways to prevent CRD are to acquire mycoplasma-free birds and provide an adequate environment and proper care. Unfortunately, birds that have had CRD can be carriers even after symptoms have long disappeared. Female carriers can pass mycoplasma to their offspring via eggs. Pens and

cages that have housed infected birds should be thoroughly disinfected prior to reuse.

Chronic Wet Feathers

Normally, ducks in good feather condition are able to stay relatively dry even in prolonged periods of wet weather. Birds with chronic wet feathers look waterlogged during wet weather or after bathing.

Causes

There are a variety of causes, including a plugged or infected oil gland, heavy infestations of mites (some kinds are difficult to see), feathers covered with fuel, sprays, or other foreign substances, and living in sloppy pens.

Treatment

If the oil gland is infected or plugged, it should be massaged gently several times daily with a warm compress, and the bird should be given an oral antibiotic as prescribed by your veterinarian.

For mites, an appropriate insecticide should be applied. In difficult cases, some people have reported success by squirting the systemic vermicide Ivomec down the throat of the patient at a dosage of $\frac{1}{10}$ cc per 4 pounds of bird. The 1-percent Ivomec solution for cattle and hogs is used. (Because the active ingredient [ivermectin] is not registered for use in birds, this product should be used only under the supervision of a veterinarian.) Birds should be treated again for mites in 4 weeks. Due to damaged feathers, ducks sometimes have to molt old feathers and grow a new set before their feathers shed water normally again.

Prevention

Keep ducks in a clean environment. Do not allow them to bathe in water that is contaminated with foreign substances, and treat them for external parasites as needed.

Coccidiosis

Coccidiosis is a major poultry disease, but unless invited by poor sanitation, it is seldom a problem in ducks. Symptoms include reduced appetite, ruffled feathers, heads drawn close to the body, and sometimes diarrhea and bloody droppings. In chronic cases, birds may grow slowly and never attain full size or production, or waste away and finally die. In severe outbreaks, large numbers of ducklings may die within a week or less. Because symptoms depend on the species of coccidia, it is recommended that birds be taken to a diagnostic lab to confirm coccidiosis outbreaks.

Cause

Coccidia are one-celled parasites that attack and destroy cells in portions of the digestive tract. There are numerous known species. These microscopic protozoa are generally present in moderate numbers where birds are raised, and they can survive in soil for more than a year. The egglike oocyst produced by coccidia can be transported on the shoes or clothing of people, in the droppings of wild birds, and by purchasing infected fowl.

Ducks generally have greater resistance to coccidiosis than chickens. However, when ducklings are overcrowded, brooded on damp litter, or kept in filthy quarters, they can suffer serious infestations. As birds mature, they normally develop immunity to "cocci."

Treatment

In general, coccidiostats manufactured for chickens and turkeys will also be effective for ducks. (One of the safest for ducks is amprolium, often sold under the names Amprol and Corid.) These preparations can be added to feed or drinking water and are usually available from feed stores and poultry supply dealers. However, the recommended dosages for chickens and turkeys should be reduced by approximately one-third to one-half for ducks, since waterfowl consume greater quantities of feed and water and overdoses of some coccidiostats can be deadly.

Prevention

The best prevention is dry, clean bedding that is turned or changed frequently to promote dryness; or better yet, wire floors in the brooding area, or at least under and around the watering containers.

Cuts and Wounds

Compared with most other animals, ducks have a normal body temperature that is feverishly high (104–109°F or 40–43°C), which protects against some infections. In reasonably sanitary surroundings, superficial scratches and abrasions usually heal naturally. However, when a duck sustains an open wound or is mauled, clinical care is needed. Prior to and after working on a wound, wash your hands thoroughly with warm, soapy water.

Treating Open Wounds

Ducks with deep or jagged cuts should usually have their feathers trimmed away from the wounds' edges. Always hold a clean piece of gauze or lintless cloth over the wound while trimming feathers to prevent bits of webbing from adhering to the exposed flesh. Wash the wound with warm water with a mild soap, and then thoroughly rinse with clear, warm water. Small pieces of shredded, loose skin that will not heal can be trimmed away.

To speed healing and to prevent infection, apply a medicated ointment (such as Neosporin) once or twice daily. To keep flies away and prevent maggots, spray a pyrethrin-based insecticide on the feathers around the wound. If open cuts are not properly cared for, infections and maggots can be problems, especially in warm weather. If the bird has multiple or severe wounds, it can help to administer a broad-spectrum antibiotic, such as Eurofloxacin Baytril (available only by prescription), orally twice a day.

Sewing Up Gaping Wounds

Stitches are required when large patches of skin have been torn loose or deep lacerations sustained. While suture needles and silk thread are preferred, a sterilized sewing needle and white thread work satisfactorily for surface suturing. Each suture or stitch should be well anchored in the skin, but not over ⅛ inch deep. Sutures should be spaced approximately ⅜ inch apart and drawn tight enough to bring the two edges of the torn flesh together without much puckering. If nonabsorbing thread is used, stitches should be snipped and pulled out with a tweezers in 4 to 5 days.

Distended Abdomen

It is normal for the abdomens of females to swell noticeably prior to and during the laying season. However, sometimes the abdomen, of either sex, enlarges to the point that a duck has difficulty walking.

Causes

There are three main causes. First, ducks that are obese (most commonly in heavyweight breeds) can accumulate large amounts of fat and their abdomens will be rather soft when palpated. Second, due to oviduct mal-functions, either yolks or fully formed eggs are dumped into the body cavity, filling it with a firm mass. Third, due to the malfunction of a major organ, the abdomen can fill with fluid, in which case the paunch is extremely heavy and tight to the touch. Various factors make birds susceptible to "water belly," including excessive amounts of calcium in grower feeds, high-protein diets, ingestion of feed containing aflatoxins, and some disease organisms.

Treatment
Birds that are obese should be put on a lower-fat diet. For internal layers, the only treatment is surgical removal of the egg mass. For "water belly," there normally is no cure. However, with special birds, symptoms can often be temporarily relieved by a high-quality diet that is aflatoxin-free and is bal-anced for all nutrients (with no more than 1 percent calcium), as well as the removal of the liquid by carefully inserting a large-gauge needle into the abdomen. However, because of the danger of puncturing internal organs, this procedure is best left to a veterinarian.

Prevention
The incidence of distended abdomen can be reduced by not letting laying birds become overly fat, providing the correct diet for birds at all stages of their lives, and providing aflatoxin-free feeds.

Duck Plague

Also known as duck virus enteritis, this deadly disease is quite conta-gious and can infect wild and domestic ducks, geese, and swans, but not chickens or turkeys. Symptoms may present as sudden death of birds that are in good flesh, listless birds that are reluctant to move, extreme thirst, diar-rhea (sometimes bloody), matted feathers on the head due to increased secretions from the eyes and nostrils, and protruding penis in dead adult males. High mortality in ducks of all ages is possible. Specimens must be sent to a diagnostic lab for positive diagnosis, and the disease must be reported to the state veterinary office.

Causes
Infections occur from direct contact with wild or domestic carriers or from water that has been contaminated by carriers of the highly contagious herpes

virus. Waterways frequented by wild waterfowl that flow into pens can transmit the virus to domestic birds.

Prevention
Never introduce new birds that are carriers. In areas where duck plague is known to occur in wild waterfowl, domestic ducks should be isolated and kept off of waterways frequented by wild birds.

Ducks kept on dry land are much less likely to contract this disease. A vaccine is available and is used by some commercial farms.

Duck Virus Hepatitis (DVH)

DVH is an acute, highly infectious disease primarily of ducklings under 8 weeks of age. It is normally limited to commercial farms with large numbers of ducks in high concentrations. The disease spreads rapidly through an infected flock, with the first deaths often occurring within 24 hours to 3 days after exposure to the virus, and an hour or less after the first symptoms appear. Mortality may be close to 100 percent in ducklings under 4 weeks of age.

Symptoms may include dull, listless birds that stop eating; watery, green diarrhea; purple-colored bills in white birds; and ducklings that retract their heads over their backs, fall on their sides, and paddle their feet. Some infected ducklings may recover rapidly, but they can continue to be carriers.

Cause
DVH is caused by a virus that can survive on contaminated equipment and in bedding for weeks. This disease is much more likely when ducklings of different ages are kept in the same building (especially if they are moved from pen to pen), and when footwear, clothing, and vehicles are not disinfected when going between flocks. Also, birds such as starlings and English sparrows, as well as rats, can transport the disease from one flock to another.

Treatment
At the first sign of symptoms, antibody therapy can reduce losses. (This is normally practical only for large operations with ready access to the antibodies.)

Prevention
Good sanitation and keeping wild birds and rodents out of brooding and growing pens are keys to preventing DVH. This disease is much less likely on

farms where each age group of ducklings is raised in a separate building that is thoroughly cleaned and disinfected after the birds are removed and prior to installing a new batch of hatchlings. Some commercial farms vaccinate their breeding stock for DVH, causing the ducks to produce antibodies that are passed through the eggs to ducklings. Ducklings can also be vaccinated.

Erysipelas

Erysipelas is relatively common in hogs and sheep, somewhat common in turkeys, and less common in geese. It is not common in ducks, but is more likely if they are raised in the same place as hogs, sheep, or turkeys with erysipelas. Erysipelas is caused by the bacterium *Erysipelothrix rhusiopathiae*, which can live in soil for a long time. If ducks are exposed to it when they are under stress, high levels of sudden death can occur. Normally, erysipelas needs to be confirmed by a diagnostic lab to distinguish it from other causes of sudden death. Because erysipelas can cause a skin disease in humans, if it is suspected that you are dealing with this disease, sick and dead animals should be handled with care and protective clothing and gloves used.

Treatment

Erysipelas needs to be treated under the supervision of a veterinarian. Recommended treatment for all the birds in a flock where some birds show symptoms includes subcutaneous or intramuscular injection of procaine penicillin along with the appropriate vaccination. Ducks showing symptoms should be penned separately from the flock and given, in addition to the above, an intramuscular injection of potassium penicillin.

Prevention

The best prevention is good sanitation and management and keeping ducks off ground known to have had infected hogs, sheep, turkeys, or geese. In areas where erysipelas is a persistent problem, birds can be vaccinated with a bacterin. (Consult your veterinarian for details.)

Eye Injuries, Foamy Eye*

Eye problems can be a result of either injuries or disease processes. Injuries often cause a bubbly foam to cover part or all of the eye. They most often occur during fighting and mating, but can also occur routinely. The incidence of foamy eye increases during times of stress and sudden weather changes.

Unfortunately, foamy eyes can also be a symptom of infectious diseases such as Chronic Respiratory Disease. However, if foamy eye appears overnight and is not accompanied by unusual nasal discharge and/or other respiratory symptoms such as coughing and wheezing, then it probably is the result of an injury or irritant.

Treatment
Usually, time and protection from further injury are all that is required. Twice-daily application of an antibiotic ophthalmic ointment to an injured eye sometimes speeds healing.

Prevention
Especially during the mating season, some drakes persistently fight with each other, and should be separated to prevent serious injury. An excessive number of drakes can greatly increase the risk of eye injuries to ducks, so the correct ratio of males to females is important. Sharp protrusions, such as projecting fencing wire and brambles, should be eliminated.

Feather Eating*

Cannibalism of this type, where birds pull out and eat one another's feathers, is most prevalent among ducklings (especially Muscovies, Mallards, and wild species) that are brooded artificially.

Causes
Feather eating is usually the result of boredom, but can also be triggered by excessively high brooding temperature, intense light, overcrowding, an unbalanced diet, or the lack of green feed.

Treatment and Prevention
At the first sign of feather eating, check brooder temperature, reduce light intensity (using blue or red bulbs often helps), and provide ducklings with sufficient space, a balanced diet, and adequate quantities of feed. As long as ducklings are kept inside, put tender green foods such as grass, lettuce, and dandelions in front of them as much as possible. As soon as it is safe, allow ducklings access to grassy pens during the day.

*Foot Problems**

Corns and calluses often develop on the bottoms of duck's feet, and normally are not a problem. However, in some cases they become infected or develop deep, bleeding cracks, causing the bird to go lame.

Causes

Foot trouble can result from bruises, cuts, splinters, or thorns in the foot pad, or from ducks spending much time on dry, hard, or sharp surfaces. However, a frequently overlooked cause is a dietary deficiency in biotin, pantothenic acid, riboflavin, or another vitamin or mineral necessary for healthy tissues. Bacterial infections (bumble foot) may also cause foot pad or joint infections, with staphylococcal bacteria commonly involved.

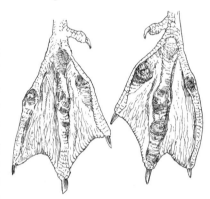

Bleeding cracks on the bottom of feet may be a result of a diet deficient in biotin, pantothenic acid, and/or riboflavin.

Treatment

If a deficient diet is suspected, supplement the rations with a vitamin premix or feedstuffs (such as brewer's dried yeast, whey, dried skim milk, or alfalfa meal) that are rich in vitamin A, biotin, pantothenic acid, and riboflavin.

If the ball of the foot is inflamed, wash the foot with warm, soapy water, disinfect with rubbing alcohol, and then remove splinter, if present. If you are sure there is a pus core, open the pad with a sharp, sterilized instrument (an X-Acto knife with a new blade or a single-edged razor blade can work here). Remove any pus or hard yellow core, disinfect the incision with iodine, and apply a medicated salve such as Neosporin.

Place the patient in a clean pen bedded daily with a layer of fresh straw (never sawdust or wood shavings) or clean towels, and provide a feed balanced for all nutrients as well as a small container of clean bathing water with 5 drops of chlorine bleach per gallon of water. Daily washing, disinfecting, and applying medicated ointment to the wound facilitate healing. For extra valuable birds, a dose of penicillin in tablet form for a minimum of 10 to 14 days seems to be helpful. In stubborn cases, a bacteriologic culture to isolate the causative bacteria and an antibiotic sensitivity test may be done by a laboratory.

Prevention

Ducks that receive a balanced diet and that are not chased over hard, sharp surfaces seldom develop foot problems. Access to pasture and bathing water reduces its occurrence.

Fowl Cholera

Fowl cholera is a highly contagious disease of both wild and domestic birds. In waterfowl, cholera often gives little or no warning, with apparently healthy birds dying suddenly. Chronic cases may be signaled by listlessness, lameness, swollen joints, diarrhea, breathing difficulties, and increased water consumption.

Waterfowl that die of an acute attack typically show little — if any — sign of the disease upon postmortem examination. In less severe cases, the liver is often streaked with light-colored areas and spotted with minute hemorrhages and gray spots of dead tissue. It is fairly common to have tiny red hemorrhages on the intestines, gizzard, and heart that are visible to the naked eye. Also, the spleen may be enlarged.

Causes

Fowl cholera is caused by the bacterium *Pasteurella multocida*, which can survive in soil and decaying carcasses for several months or longer. It can be spread by wild birds, rodents, and scavengers, or from ducks pecking at infected dead birds. Although cholera can occur any time of year, it thrives best in a damp, cool environment.

Treatment

Recommended treatment is one of the following sulfa drugs: sulfaquinoxaline sodium at the rate of 0.04 percent in drinking water, or 0.1 percent in feed for 2 or 3 days; sulfamethazine at 0.4 percent in feed for 3 to 5 days or sulfamethazine sodium, 12.5 percent solution at 30 mL per gallon of drinking water; or sulfamerazine sodium at 0.5 percent in feed for 5 to 7 days. Sulfa drugs must be used with caution, particularly with breeding stock, as they can be toxic. High levels of antibiotics such as tetracycline are sometimes used in the feed or injected under the skin. For the small-flock owner, the most practical treatment is usually adding easy-to-use prepared medications to the drinking water, such as Salsbury Sulquin. For sulfa-resistant infections, penicillin given intramuscularly is often effective.

Prevention

Sound sanitation practices are the best prevention. Water containers should be placed over wire-covered platforms and waterers frequently cleaned and occasionally disinfected with sodium hypochlorite (common bleach used at 4 ounces per gallon of water as a disinfectant) or an approved livestock sanitizer. Eliminate stagnant mudholes in the duck yard and burn or deeply bury all carcasses of dead birds and animals. In localities with a history of cholera, pasture rotation and vaccination using commercially available bacterins according to the manufacturer's recommendations may be necessary.

Frostbite

During freezing weather, look for birds that have their feet frozen to the ground or ice. Also watch for ducks limping when forced to walk; swollen and/or red feet that feel hot to the touch; and tissue that is sloughing off. Unfortunately, frostbite is often not detected until lameness, gangrene, or discoloration occurs.

Cause

Prolonged exposure of the feet (or bare facial skin of Muscovies) to extreme cold can result in freezing of tissues.

Treatment

When waterfowl are found with their feet frozen to ice or the ground, pour warm water that is 90 to 105°F — no hotter — over the frozen parts until they are freed. Then, rapidly warm the frostbitten feet in a water bath (105 – 108°F) for 15 to 20 minutes, and give the patient lukewarm drinking water. Do not rub the affected parts. If gangrene sets in, the frozen areas may eventually drop off or may need to be amputated and treated as an open wound. The oral administration of antibiotics such as penicillin and Terramycin to birds with severe frostbite reduces the chance of infection.

Prevention

Unless waterfowl have access to a large body of open water, ducks of all breeds should be enclosed in a yard or shed with a thick layer of bedding, and provided protection from wind when temperatures fall below 20°F.

Hardware Disease and Esophagus Impaction*

Wire, nails, strings, and other objects may seem small and innocuous, but they can wreak havoc if ingested by your ducks. With hardware disease, birds may slowly lose weight, stop eating, or sit around with eyes partially closed, apparently in severe pain. When a postmortem is performed, the hardware is often found lodged in the esophagus or gizzard. Frequently, the foreign object will penetrate the tissue wall, causing peritonitis. With an esophagus impaction, a lump, caused by an obstruction, is often visible in the lower neck.

Causes

Nails, bits of wire, pieces of string, blades of tough grass, excessive quantities of gravel, or other hard-to-digest objects are sometimes swallowed by ducks. When ingested, these objects may puncture or impact some portion of the upper digestive tract.

Treatment

There is no simple remedy for hardware disease. If you deem the bird sufficiently valuable and can locate a veterinarian willing to perform the surgery, it is possible, but risky, to remove the hardware from the gizzard.

For an esophagus impaction, the wad of material can often be kneaded loose by gently massaging the compaction from the outside, working it back up and out of the mouth. (Tubing the bird with warm water and then holding it upside down while massaging the compaction is often helpful.) If relief cannot be achieved by external methods, the blockage may have to be removed surgically.

To prepare for this operation, pluck the feathers directly over the impaction a few at a time until an area approximately 1½ inches in diameter is exposed. After washing your hands with soap and warm water for several minutes, drench the plucked patch with peroxide, gently tighten the skin by stretching it between the thumb and index finger, and make a shallow, inch-long incision through the skin with a sterilized (boiled for 3 minutes) knife, such as an X-Acto knife with a new blade. A second incision is made through the wall of the esophagus.

Using your finger or a sterilized, blunt instrument, remove the troublesome material from the esophagus and then rinse with clean, warm water. Using a fine needle and gut suture material (gut must be used so that it will dissolve), draw the incised edges of the esophagus together with three or four single stitches that are tied off separately. The outer cut can be sewn in a similar manner, except that silk thread should be used. When finished, wash the incision with peroxide, apply an antibiotic ointment, force-feed several capsules of cod-liver oil, and provide drinking water, but no feed, for 24 hours. Thereafter, supply small quantities of easily digested greens, such as lettuce, and pellets several times daily until the stitches are removed. Apply an antibiotic ointment daily until signs of inflammation have dissipated. The outer sutures can be taken out after about a week.

Prevention

Never leave nails, wire, or string where birds can reach them. When hardware is used in an area to which birds have access, a strong magnet comes in handy for picking up dropped wire and nails.

It is inadvisable to place large quantities of grit in troughs with feed. If ducks have been without grit for some time, give only 1 teaspoonful per bird every other day for a week before giving grit free-choice. I have seen situations in which it appeared that ducks that had been deprived of grit ate such large quantities of sand or pea-sized gravel once it was available that their gizzards became impacted and the birds starved to death because feed could not pass through to the lower intestines.

Lameness*

Lameness can have many causes, including dislocated hip, sprains of leg or foot joints, infections (cuts, abrasions, splinters), muscle damage, pinched nerves, dietary deficiencies (especially of niacin, biotin, and other B vitamins), or a calcium:phosphorus imbalance (common when high-calcium laying rations are fed to immature birds). Also important, but less common, are inherited leg weaknesses. Injuries can happen when ducks are chased, mauled by predators, handled improperly, startled at night (causing them to panic and run blindly into objects and walls), stepped on by larger animals, punctured by brambles and other sharp objects, or bruised by extended time on hard or rough surfaces.

Symptoms

Lameness can have sudden onset or develop gradually over days. Sprains are normally accompanied by swelling in the injured joint. Infections can also cause inflammation in the joints as well as other parts of the legs or feet. Pinched nerves can cause partial or complete paralysis of the leg. Leg problems associated with dietary deficiencies, calcium:phosphorus imbalance, or inherited weaknesses are usually signaled by legs that tremble when the duck stands still, give out after the bird walks or runs a relatively short distance, are bowed, or are twisted out at the hock joint. (Also, see Spraddled Legs.)

Treatment

If otherwise healthy, ducks with injured legs will normally recover if kept quiet in a clean pen and provided easy access to food that supplies a balanced diet and drinking water to which a good poultry vitamin mix has been added. Clean swimming water can help a duck with a serious leg injury to recover.

When a deficient diet is the cause, take prompt action. Mixing a vitamin/mineral supplement in the drinking water or with the feed (per manufacturer's recommendation), or the feeding of 2 to 3 cups of brewer's dried yeast per 10 pounds of feed, will often correct the problem. If you have been feeding immature birds a high-calcium laying ration, immediately switch to a feed that contains a *maximum* of 1 percent calcium and a phosphorus:calcium ratio of 1:1 to 1.0:1.5.

Prevention

Select breeding birds that display strong limbs from hatching to maturity. Don't catch or carry ducks of any age by their legs or run them across rough ground or over equipment such as water and feed troughs. Feed ducklings a diet fortified with niacin and vitamins D and A, and with the proper phosphorus:calcium ratio, in the range of 1:1 to 1.0:1.5. Young birds should not be fed laying rations, as they have too much calcium and an improper phosphorus:calcium ratio, which can cause serious problems.

*Maggots**

When open wounds are left untreated, especially during warm weather, blow flies may be attracted to the wound sites and may then lay eggs on sores. In a short time, the eggs hatch into maggots and feed on surrounding tissue. During the breeding season, the backs of females can be lacerated as the males tread them. These lacerations can be prime sites for maggots.

Treatment

Prevention is the best treatment. Birds with wounds should have an antibiotic ointment applied daily until healed. As a precaution, some people give birds with wounds an oral dose of ¹⁄₁₀ of a cc of 1 percent (injectable) Ivomec per 4 pounds of bird.

Ivermectin gives protection against blow flies for approximately 2 weeks. A word of caution: Ivermectin is not registered for use in birds, and therefore should be used only under the supervision of a veterinarian. In situations where maggots are present, one technique is to spray a small amount of car-starting fluid onto the maggots. Some will fall off and the others can be removed with a tweezers. Hydrogen peroxide can then be squirted with an eye dropper into crevices, which will cause embedded maggots to back out, facilitating removal. The wound should be treated daily with an antiseptic spray or ointment and checked for missed maggots until completely healed.

Prevention

Ducks who develop phallus prostration, prolapsed oviducts, or open wounds should be cared for promptly or destroyed to prevent a maggot infestation. Birds with wounds or open sores should have a pyrethrin-based insecticide sprayed daily on the feathers around the wound (cover the wound with a clean cloth to protect from the spray).

Niacin Deficiency*

One of the most frequent problems that our farm gets calls for concerns young ducks suffering from niacin deficiency. Birds develop weak or bowed legs, and often show stunted growth and enlarged hocks. In mild cases, a classic symptom is that some of the birds are of normal size and have strong legs while others are undersized and show leg deformities. (Rickets, sometimes confused with niacin-deficiency symptoms, is caused by a vitamin D_3 deficiency or a calcium and/or phosphorus deficiency or imbalance.)

Cause

This deficiency is caused by ducklings consuming a diet low in usable niacin. Most chick starters are deficient in niacin. It is also important to remember that most of the niacin in plants is not available to ducklings.

Treatment

Ducklings exhibiting symptoms of a niacin deficiency can often be cured by the immediate addition of 100 to 150 mg of niacin per gallon of drinking water until they are 8 to 10 weeks old. (Niacin in tablets or powder is available at drugstores). If niacin deficiencies are not treated in time, ducklings can be permanently stunted and/or crippled.

Prevention

Ducklings must be fed a diet providing 35 mg of available niacin per pound of feed from 0 to 2 weeks of age and 30 mg from 2 to 10 weeks of age. If regular chick starter is used, add niacin in the drinking water at the rate of 100 to 150 mg per gallon from 0 to 10 weeks of age. When allowed to forage in areas with an abundant supply of insects, ducklings seldom experience a niacin deficiency.

Nutritional Deficiencies*

There are many possible symptoms; the most common ones in ducklings include wide variation in body size in flocks with birds of the same age, stunted growth, retarded feather development, weak and/or deformed legs, and birds with reduced resistance to disease or parasites. Adult ducks that are undernourished produce poorly, often have rough-looking feathers, may be thin, and are susceptible to disease and parasites.

Causes

Malnutrition occurs most frequently when ducks are raised in buildings or grassless yards and fed nothing but grains or inadequate chicken rations. Unfortunately, mistakes are sometimes made in the manufacturing of feeds, and birds can be malnourished even though you are feeding them a supposedly adequate ration. Other causes include heavy parasite infestations, or extremely hot weather, which reduces feed intake.

Treatment and Prevention

Provide an ample quantity of food that supplies a balanced diet. As temperatures go up, feed consumption goes down, so in hot weather the nutritional density of feed must be higher for adequate nutrition. If you are trying to save money by using the cheapest feed available, keep in mind that per unit of nutrition, higher-priced feeds are often the best deal.

Omphalitis

When ducklings hatch, their navels are sometimes infected by micro-organisms that cause omphalitis. Trouble can be signaled by inflamed navels or abdomens that are abnormally distended and mushy-feeling. If not observed closely, ducklings with omphalitis can appear fairly normal until a short time before expiring. Infected birds in advanced stages usually huddle close to the heat source and move about reluctantly. Mortality, which may be light or heavy, invariably takes place between the 2nd and 6th days after hatching.

Cause
It is inevitable that bacteria are present in nests and incubators, but poor sanitation increases the chance of omphalitis. Excessive humidity compounds the problem by slowing down the normal healing process of the navel and providing an ideal habitat for the bacteria.

Treatment and Prevention
Prevention is the only effective therapy. Providing an adequate number of nests that are generously furnished with clean nesting material is where prevention begins. (Granted, some ducks refuse to use nests!) Soiled hatching eggs should be washed with clean, lukewarm water to which an appropriate sanitizer, such as Germex, has been added. Finally, the incubator and hatcher must be kept clean and disinfected after each hatch. (Whenever possible, incubate eggs in one machine and then hatch them in a separate hatcher, thoroughly disinfecting after each hatch.) Duck eggs should be candled on the 6th, 12th, 19th, and 24th days of incubation to avoid "blow-outs" that spew millions of virulent bacteria into the incubator or hatcher.

Oviduct, Eversion of the*

While attempting to lay eggs, females will occasionally expel a portion of their oviduct. A duck with this problem is easily identified by the expelled portions of the oviduct protruding from her vent.

Causes
Possible causes include obesity, premature egg production, oversized eggs, excessive mating, and prolonged egg production. My experience indicates that the first three causes are by far the most significant.

Treatment

An ailing duck can be saved only if she is discovered relatively soon after the oviduct is dislodged and prompt action is taken. Carefully catch the bird, doing what you can to keep the oviduct clean. Gently wash the protruding oviduct with clean, lukewarm water to remove dirt and feces. Mineral oil can be applied to reduce drying of the tissues. Normally, if the oviduct is pushed back into place, it will come back out unless a few purse-string sutures are placed in the vent. (This continuous stitch allows eggs and feces to pass without tearing out the sutures, and is best installed by a veterinarian.)

During the recovery period, the patient should be kept in a warm, dry, clean pen away from all males (a female companion may help keep her calm) and fed a non-layer feed to discourage laying. Once the abdominal muscles have had time to strengthen, sutures can be removed (usually in 8 to 12 days). Full recovery is aided by isolation from drakes and discouraging laying for at least 3 months.

Prevention

Do not push young ducks into laying before they are 18 to 20 weeks old, and make sure females are in good flesh, but not overly fat, at the beginning of and throughout the laying season. If excessive mating is a problem, drakes with overabundant libido can be penned separately and put with the females for several hours twice a week.

Parasites, External

It is normal for healthy ducks that live outside to have some lice. If overdone or used recklessly, insecticides can cause more harm than good. Also, insecticides that come in contact with the vent can cause temporary sterility.

The first indication that a bird has external parasites can be repeated scratching of the head and neck with the feet. If you look closely under a good light, body, head, and neck lice usually can be seen with the naked eye by parting the feathers of the head, neck, and around the vent and oil gland. By holding an open wing up to a light source, such as the sun, wing lice — if present — are visible as dark lines ⅛ to 3⁄16 inch long in the webbing of the secondary and primary flight feathers. Mites are often overlooked because of their tiny size, plus they may be on the victim at nighttime only. In severe infestations, mites sometimes are visible swarming over the surface of feathers or on the skin when feathers are parted and exposed to sunlight. After handling an infested bird, the miniature ticklike mites may also be seen and felt on your hands and arms.

Lice-infested ducks may have retarded growth, lose weight, lay poorly, and be "on edge" from the irritation of the lice. Heavy infestations of mites cause slow growth, anemia, weight loss, deserted nests, and even death.

Causes

When waterfowl are raised in a relatively clean environment and have access to bathing water, lice and mites do not often multiply enough to be troublesome. (In some localities, external parasites are much more prevalent than in others.) However, ducklings that are hatched or brooded naturally or adult ducks that do not have access to swimming water can harbor harmful infestations of external parasites.

In general, lice are small, flat, yellowish tan insects that normally live their entire life on the host bird. The most common species found on ducks include the wing louse, head and neck louse, and body louse. Mites are blood suckers for the most part. Some species, such as the feather mite and de-pluming mite, stay on the birds most of the time, while the red mite normally is on the birds only while feeding.

Treatment

Various treatments include olive oil, pulverized dried tobacco leaves, organic apple cider diluted 50 percent with water, or commercial insecticidal preparations such as Sevin, Malathian, or Mange & Lice Control. To be effective, these products need to be worked into the feathers of the head, neck, wings, upper tail, back, and vent. When dusting or spraying with an insecticide, be extremely careful not to contaminate water or feed. In case of heavy mite infestations, buildings, nests, and roosting areas should be cleaned, disinfected with an approved disinfectant, and then sprayed with a product such as Mange & Lice Control.

Prevention

To keep external parasites in check, provide sanitary living conditions, supply bathing water when possible, and treat birds before lice or mites are numerous enough to be harmful. Turkey and chicken hens used as foster mothers should always be treated for lice and mites before their maternal chores begin.

Parasites, Internal

Healthy waterfowl naturally have reasonably good immunity to most internal parasites. An infestation of worms causes retarded growth, lowered feed conversion, reduced egg production, and increased susceptibility to disease. (In the case of gapeworms, birds often attempt to clear their windpipes by vigorously shaking their heads and coughing.) In severe cases, the above symptoms are accompanied by weight loss, weakness, diarrhea, and, if not treated promptly, eventually death. Upon postmortem examination, most species of worms can be found in their appropriate habitat, if they are present in significant numbers.

Causes
Worms are primarily a problem when ducks have access to stagnant water, crowded ponds, or small streams, or when forced to survive in a filthy environment. A variety of worms occur in poultry, including large roundworm, cecal worm, capillary worm, gizzard worm, gapeworm, and tapeworm.

Treatment
Treat ducks for internal parasites only if they actually have an infestation. The indiscriminate use of wormers over a period of time can reduce the natural immunity of the birds. The best way to know if your birds need treatment is to take a fecal sample to your veterinarian for evaluation.

Poultry worm medications to add to drinking water are readily available and easy to use. Some wormers work on only one species of parasite, while others are effective against several. Check with your feed or poultry supply dealer for available brands and follow instructions carefully. For a persistent problem, a carefully planned worming schedule as outlined by the manufacturer will be needed to eradicate the parasites. Deworming ducks that are laying can negatively affect egg production and fertility. Some people have suggested diatomaceous earth or a pinch of chewing tobacco pushed down each bird's gullet as alternative remedies.

Prevention
Good sanitation practices, adequate diet, and the treatment and quarantine of all new arrivals to the farm are key to preventing heavy parasite infestations. Watering vessels should be placed on wire- or slat-covered platforms.

Pasteurella anatipestifer *Disease*

Also known as New Duck Disease, Duck Septicemia, and Infectious Serositis, Anatipestifer is most common in 2- to 9-week-old ducklings raised in dense concentration with poor sanitation and management. Symptoms in ducklings are similar to those in chickens with Newcastle disease: discharges from the nostrils and eyes, loss of appetite, green diarrhea, overall listless appearance, coughing, sneezing, head and neck tremors, and a stumbling gait. Ducklings that survive may develop swollen hock joints.

Cause
The bacterium *Pasteurella anatipestifer* is the cause; it has an incubation period of 1 to 5 days, with the first mortalities from a few hours to a week after the first visible symptoms. Older birds are typically more resistant. Mortality can range from less than 10 percent to more than 75 percent. Good management practices reduce the mortality rate. *P. anatipestifer* is probably spread through the drinking water and inhalation of infected airborne particles.

Treatment
Birds showing severe symptoms normally do not respond to treatment. If caught early, a single injection of a combination of penicillin and strepto-mycin or sulfaquinoxaline in the feed or water normally reduces losses. (Streptomycin is toxic to some strains of birds, so it is a good idea to inject only one or two birds and then observe for 30 minutes. If they are sensitive, they will get droopy and stagger.) A good poultry vitamin mix added to the drinking water may help boost their immune system.

Prevention
Good hygiene and management are the keys to prevention. New stock should be quarantined for 3 months, wild waterfowl kept away, and visitors kept out of pens. In areas where *P. anatipestifer* is common, some commercial producers use a bacterin vaccine.

*Phallus Prostration**

The penis, a 1½-inch-long (or longer) organ, protrudes from the bird's vent. Frequently, a drake with this disorder will be seen repeatedly shaking his tail from side to side as he attempts to retract the decommissioned organ.

Causes

Wild drakes normally pair off with a single duck and are sexually active for a relatively short period each spring. Under domestication, males often become sexually active at 3½ to 5 months of age, have multiple mates, and may breed the year round. Some drakes lose the ability to retract their penises, possibly caused by this unnaturally long and active mating season. (Some people have suggested that a genetic weakness is involved, but the evidence does not support this claim.)

Treatment

Drakes with phallus prostration should be immediately isolated in a clean pen to prevent other ducks from damaging the penis by nipping it. If the bird is valuable and his problem is discovered before the penis has become infected or dried out, recovery is possible. The organ should be washed with clean, warm water and then disinfected and treated with a medicated oint-ment. You can try pushing the penis back into place, but — as often as not — it will pop back out in a short time. Apply the ointment daily until he is fully recovered, which may require several weeks or longer. Drakes should not be allowed to mate for at least 3 months.

Often, the end of the penis dries up and falls off. Because the semen exits from the base and travels down a groove that spirals around the penis, a drake can still be fertile even if a portion of the penis sloughs off (although fertility usually does not return for 6 to 12 months).

Prevention

The best safeguard is to keep backup drakes for breeding purposes.

*Poisoning from Medication**

Medication poisoning occurs either because the drug used is toxic to ducks or it was given in the wrong dosage. Symptoms vary depending on the age of the birds and the particular medication. Ducks may get sleepy, stagger, show signs of neck paralysis, lose their appetites, become weak, have stunted growth, or die suddenly.

Causes

In proportion to their size, ducklings consume more feed and water than chicks or turkey poults. Therefore, when given medications in their feed or

water at the rate designated for chicks and poults, they are consuming overdoses. Also, some medications may be toxic to waterfowl but not landfowl.
One prominent example is Ren-O-Sal, a medication commonly used for
chickens that is deadly to ducklings. However, most modern medications, if
given in the correct dosages for ducks, are safe.

Treatment

At the first sign of any of the above symptoms, discontinue the medication.
If medicated feeds must be used, keep an eye out for symptoms of possible
medication poisoning.

Prevention

Do *not* use Ren-O-Sal with ducks! Medications that showed no negative
effects in feed trials with ducks include amprolium, sulfaquinoxaline,
Zoamix, bacitracin, chlor-tetracycline, dinitro-ortho-toluamide, ethopabate,
furazolidone, novobiocin, and neomycin. In normal circumstances, do not
use medicated feeds unless there is a particular need or it's the only feed
available.

*Poisoning from Plants and Other Substances**

Besides spoiled food and certain medications, other organic and inor-
ganic substances can be poisonous to poultry if ingested in sufficient quantities. Diagnostic signs vary depending on the poison and the quantity
ingested. In general, common symptoms at low or nonfatal levels include
retarded growth, droopy appearance, and unsteadiness. At high levels, birds
may go into convulsions, fling their heads from side to side (apparently
attempting to regurgitate the contents of their esophagus), or die suddenly.
Due to the free gossypol content of untreated cottonseed meal, excessive
feeding (more than 10 percent of the ration) of this protein supplement can
result in suppressed growth, reduced egg production, and discolored egg
yolks and albumin.

Causes

Some common materials that are known to be toxic to ducks include commercial fertilizers; salt; lead (from birds picking up lead pellets or nibbling
on leaded paint); herbicides; pesticides; baits for rodents, slugs, and snails;
leguminous plants and their raw seeds; cottonseed meal; and leaves of

tobacco and rhubarb. Other plants that are suspected of causing illness or death in ducks include foxglove, potato vines, potatoes that have turned green, and eggplant leaves. This is only a partial list; there are many poisonous plants in various parts of the world.

Prevention
Whenever using poisonous baits of any kind, locate them out of the reach of livestock. If commercial fertilizer is applied, do not spill or store it where ducks can get to it, and always make certain that the granules are dissolved for at least a week by rain or irrigation before birds are permitted on fertilized pasture. Buildings and equipment with peeling lead-base paint must be cleaned up or made inaccessible to ducks. Birds should not be allowed to eat sprayed or poisonous plants, salt, or icy slush resulting from snowy driveways or sidewalks being salted in winter. Common vetch seed and cottonseed meal should not be included in a ration for ducks at more than 5 percent. (For breeding ducks, the use of cottonseed meal is questionable.) Raw soybean meal should not be used and alfalfa meal should not make up more than 5 percent of a ration. Even "safe" herbicide sprays should be used with great caution around ducks since they can cause decreased fertility in breeding birds.

*Sinus Infection**

See also Chronic Respiratory Disease. Normally, the first noticeable symptom is a swollen sinus (between the bill and the eye), but nasal discharge, foamy eyes, and throat rattles may also be present.

Causes
Various microorganisms can cause sinus infections in ducks, but *Mycoplasma gallisepticum* is culpable in a majority of cases. The infection often lies dormant until a bird is stressed by factors such as shipping, moving to a new pen, sudden feed changes, drastic weather changes, or inadequate diet.

Treatment
At the first sign of a sinus infection, the affected bird should be isolated to minimize infecting other birds. There are many strains of mycoplasma and they vary in their sensitivity to different antibiotics. Tylosin (sold as Tylan) often is by far the most effective, followed by erythromycin (Gallimycin) and tetracyclines (Aureomycin and Terramycin).

Prevention

The best way to prevent sinus infections is to acquire mycoplasma-free stock. Unfortunately, birds that have had sinus infections can be carriers even after symptoms have long since disappeared.

Spraddled Legs*

Ducks may have legs that slide out from under them as they attempt to walk. In acute cases, the legs may protrude at right angles from the body. It is usually a minor problem if quickly attended to.

Causes

Spraddled legs can usually be traced to smooth bottoms in incubator trays, brooder floors, or pens where birds have poor footing.

Treatment

Provide floors with good footing. Tying a short piece of yarn between the patient's legs for several days will often cure spraddled legs. When hobbles are used, it is essential that they are not tied so tightly that blood circulation is restricted.

Prevention

Never put ducks (especially heavy breeds) in pens or boxes with slick surfaces.

Spraddled legs (left) can often be remedied if legs are hobbled (right) with a piece of yarn.

Staphylococcosis

In ducks, the most common places for staph infections are the joints of the feet and legs. In the mild form, birds show lameness, an unsteady gait, stand or sit by themselves, move about reluctantly, lose weight, and, if not treated promptly, eventually die. In severe cases, the above symptoms can be accompanied by hot, swollen joints, diarrhea, and sudden death. Postmortem exams generally show congested and swollen livers, spleens, and kidneys.

Causes

The organism that causes staph infections, *Staphylococcus aureus*, can be found in most flocks of poultry, but does not seem to be a serious threat unless ducks are in a run-down condition physically, due to poor nutrition, parasites, injuries, or from being kept in grossly unsanitary quarters.

Treatment

Depending on the strain of staph, various antibiotics can be effective, including: Gallimycin, Terramycin, novobiocin, and penicillin. (Prescription antibiotics can be obtained from your veterinarian.)

Prevention

When ducks are kept in reasonably sanitary conditions and adequately fed, staph is usually not a major problem. In brooding pens, fresh bedding (we use coarse cedar sawdust) should be sprinkled on top before pens get greasy. During rainy weather, the duck yard should not be allowed to deteriorate into filthy, mud-covered cesspools. Birds should not be allowed to walk on sharp objects that may puncture their feet, thus inviting this infection.

*Sticky Eye**

In this condition, the feathers around the eyes are matted, and in severe cases the eyelids stick together.

Causes

Under domestication, waterfowl are sometimes raised on diets deficient in vitamins and minerals such as vitamin A, pantothenic acid, and biotin. They also may be supplied water in shallow containers that do not permit them to rinse their eyes. The result is that ducks raised in confinement are sometimes susceptible to sticky eye. Normally, sticky eye is not a health problem if it is simply a matter of insufficient bathing water.

Treatment

If the cause is not an infection or dietary deficiency, adequate bathing water normally takes care of the problem. (Some birds need a large pool before they will adequately clean their faces.)

Prevention

Sticky eye can be kept to a minimum by supplying an adequate diet and clean bathing water.

Streptococcosis

Under most circumstances, streptococcosis is uncommon in ducks. The strep infection found in ducks is caused by the bacterium *Streptococcus gallinarum* and can be introduced into flocks by carrier birds. This disease is difficult to diagnose, and is probably occasionally mistaken for fowl cholera due to similar symptoms, including the sudden death of birds that appear to be in good health. The liver and lungs often show congestion and enlargement. Streptococcus must be isolated in the laboratory for positive identification.

Treatment and Prevention

Treatment is difficult since birds often die before infection is suspected. Antibiotics that are useful against gram-positive organisms (check with your veterinarian) have been suggested as a preventive measure and as a treatment for infected fowl when used at the highest recommended level. Acquiring healthy stock that has been raised in a clean environment and providing sanitary living conditions seems to be the best prevention.

Sudden Death

In this case, a duck is found dead that a short while earlier appeared normal and healthy.

Causes

There are a host of causes including heart attack, choking, hemorrhaging, blood clots, broken egg in oviduct, suffocation, exposure to toxic gases, consumption of toxic substances, being kicked or stepped on by a larger animal, attack by a predator (puncture wounds can be almost impossible to detect on heavily feathered portions of the body), heat stroke, seizure, broken neck, virulent disease organisms, and drowning. If ducklings are held so tightly that their bodies cannot expand and contract for breathing, they can be suffocated quickly.

Prevention

Some sudden deaths cannot be prevented. The rate of sudden deaths can be reduced by a proper environment, including plenty of shade and cool drinking water during hot weather; protection from large animals (that might step on them) and predators; avoiding situations that terrify ducks or cause them to panic; keeping the correct ratio of males to females (drakes may gang up on females and drown them while attempting to mate on water); and always using proper methods for catching and holding ducks. Good sanitation and management reduce the chances of infectious diseases that cause sudden death.

Wing Disorders*

There are three main wing disorders: twisted wing (the wing tip sticks out from the body when the wing is folded), split wing (there is a gap between feathers in the wing), and lazy wing (the tip of the wing hangs down along the side of the body). These are primarily an aesthetic concern, and seldom affect productivity.

Causes

The most common cause is a combination of improper diet and insufficient exercise. Improper diet can include excessive or inadequate protein, vitamins, and minerals. In general, diets containing more than 16 percent protein after ducklings are 2 weeks old greatly increase wing disorders. Ducklings raised outside with access to swimming water and an appropriate diet seldom have wing disorders. Other causes include excessively high brooding temperatures, feather eating, broken wing feathers, narrow feathers (common in Runners), injuries, and genetic defects.

Treatment

When treated before the bones of the wing harden into position, it is often possible to repair twisted wings of ducklings by manually folding the feathered limb in the correct position and taping it shut with ½-inch masking tape. Because the wing will become stiff from nonuse, the tape should be removed after 10 to 14 days, and if need be, retaped after half a day of exercise. For mature birds that cannot be rehabilitated, clipping the flight feathers improves their appearance.

Prevention

Wing disorders can be greatly reduced by feeding balanced diets with no more than 18 percent protein from 0

Twisted (also called slipped) wings can sometimes be fixed by temporarily taping them in position.

to 2 weeks and 15 to 16 percent protein up to 10 weeks. Ducks should be allowed outside to get exercise as soon as it is safe.

Precautions when Using Drugs and Pesticides

When used according to the instructions and along with good management practices, the occasional use of drugs and pesticides can be useful in maintaining the health and productivity of poultry flocks. However, if overused or misused, these aids pose health hazards to both animals and humans. Always follow directions carefully and use only the dosages recommended for the specific problem. Don't fall into the trap of thinking that if one dose is good, then a double dose must be better. To keep from poisoning you and your family or customers with potentially dangerous drugs, follow recommended withdrawal periods when treating meat- or egg-producing birds. And last but not least, store all drugs and pesticides in their original containers in a dry, clean location out of the reach of children and animals.

Pullorum-Typhoid Blood Testing

Pullorum and fowl typhoid are two types of highly contagious Salmonella infections that can be passed from breeding stock to offspring through hatching eggs or by infected birds coming in contact with healthy birds. Since these diseases are easily transported from one locality to another, some states (and most countries) require that all adult poultry crossing their borders, either for breeding stock or for exhibiting in a show, must be blood tested for pullorum-typhoid and certified clean. For transporting or shipping day-old poultry, the parent stock must be certified clean.

For the small-flock owner, blood testing is not usually required (although it's not a bad idea and is relatively inexpensive in states that cooperate with the National Poultry Improvement Plan) unless birds are to be exhibited or sold in states requiring a health permit. If you acquire your stock from breeders who do not annually blood test their birds, there is some risk of an outbreak.

For waterfowl and chickens, the pullorum-typhoid test is normally performed by extracting a small amount of blood from under the bird's wing and mixing it with a drop of antigen on a light table. To be valid, blood tests must be performed by a licensed technician. For more information, contact the veterinary office of your state or province Department of Agriculture.

Postmortem Examination

Most of us who raise poultry have experienced the disappointment of finding an expired bird that was in good flesh and showed no outward signs of disease or attack from a predator. When this occurs, you can: 1) dispose of the carcass and never know what caused the bird's demise; 2) take the fowl to a diagnostic laboratory or veterinarian for diagnosis; or 3) perform a postmortem yourself and see if the problem can be located.

The first of the three choices is probably the most common but definitely the least desirable. The second is most desirable but least common and sometimes impractical, except when a serious or contagious disease is suspected. While laypersons cannot approach the proficiency of trained diagnostic specialists, the third alternative can be useful in identifying many problems to avert further mortality or lowered production.

Equipment Needed

The average household contains the few tools needed to perform a basic postmortem examination. A small, sharp knife (which afterwards should not be used on human food) is needed for opening the bird. A pair of scissors is useful for incising the trachea (windpipe), esophagus, intestines, and ceca. A magnifying glass is helpful in searching for internal parasites such as hard-to-see capillary and cecal worms. The workbench should be covered with newspapers and located in a well-lit area. *CAUTION:* After a postmortem, all tools must be sterilized and the work area and your hands disinfected to guard against spreading infections and disease to humans and livestock.

Procedure for a Basic Postmortem

When examining waterfowl, I have adapted the following procedure, although the sequence of the examination may vary. If an esophagus or gizzard impaction is suspected, these organs are inspected first; if worms are a prime candidate, then the intestines are examined first, and so on.

1. Cut and spread open the bird from one corner of the mouth to the vent.

2. Scan the body cavity and organs for hemorrhages and tumors.

3. In mature females examine the body cavity and oviduct for abnormalities such as internally laid eggs, blocked oviduct, or obesity.

4. Examine the liver for discoloration, tiny light spots, light streaks, dark hemorrhagic areas, hardness, or yellow coating of waxy substance.

5. Cut open the gizzard and check for a hard mass of string or other material, eroded inner lining, or serious damage. Peel off the horny inner lining and look for uncharacteristic bumps, which may indicate gizzard worms.

6. Open the esophagus and look for blockage or injury.

7. Examine the main organs (liver, heart, spleen, lungs, kidneys, and ovaries) for obvious deformities and discoloration.

8. Slit open the small and large intestines and ceca, checking for worms, blood, inflamed linings, hemorrhages, and yellow, cheesy nodules.

9. Split the trachea lengthwise and inspect for blockage, gapeworms, blood, excess mucus, and cheesy material.

10. When finished with the examination, clean up and dispose of the carcass and all debris, disinfect the tools, and thoroughly wash your hands and arms with soap and warm water.

11. To maximize the effectiveness of the postmortem examination, you can write up a short report of your findings for future reference.

EIGHTEEN

SHOW TIME

North American poultry breeders have been displaying their fowl at public exhibits for over 150 years. By digging into the annals of poultry history, we find that the first exclusive poultry show was held on November 14, 1849, at the Public Gardens in Boston, Massachusetts. The Boston Poultry Exposition continues as an annual event.

Poultry shows serve a number of useful functions. In addition to providing a meeting place for old and new friends who share a common interest, they allow the public to see and enjoy the breeders' skills and provide the opportunity to compare one's stock and ideas with those of other breeders. Shows can also be excellent places to advertise and sell your birds, and many people enjoy the competition of exhibiting their ducks.

Kinds of Shows

Shows come in many sizes and shapes. There are county, state, provincial, club, and national shows, ranging in size from a few dozen birds to several thousand — the largest having topped 10,000 entries. At county and state fairs, the open-class division can be entered by anyone, whereas the 4-H and FFA (Future Farmers of America) divisions are limited to members. Club shows are usually divided into youth division (for anyone 18 years and younger) and open class (for exhibitors of all ages).

Although shows were originally held primarily in the fall, today they are held in all seasons of the year in some regions. For information on show dates and locations, call local fairgrounds, check with local poultry clubs, and subscribe to the *Poultry Press* and *Feather Fancier,* monthly newspapers that carry advertisements for most club shows. (See Appendix H, page 302.)

Open Class

In the open-class division, most ducks are judged in their show cages as the judge walks along the aisle. Unlike with chickens, judges often do not take the birds out of their cages in order to minimize the possibility of injuries. At some shows, Runners are judged in a show ring and the owner or show assistants carry the birds from their cages to the show ring at the appointed time.

4-H, FFA, and Youth Showmanship Class

In most 4-H and FFA shows, birds are carried by their owners or a helper to a show "bench" where they are presented to the judge for evaluation. In conformation classes, the birds are judged by the prevailing breed standard. In showmanship classes, the primary emphasis is on the knowledge and appearance of the owner and the behavior of the bird on the show bench. For showmanship, most fairs do not discriminate against crossbreds or mediocre purebreds. County extension agents normally have material available on preparing birds and exhibitors for showmanship classes.

Who Can Show?

The majority of poultry shows are open to everyone who wants to enter their birds. Belonging to a local or national organization is not usually a prerequisite for exhibiting your ducks, although some county and state fairs limit entrants to residents. To help cover expenses, there is a reasonable entry fee. In some states, any fowl that is shown from out of state must be blood tested for pullorum-typhoid; a few states require the test for in-state birds as well. Premium lists spell out such requirements and the show secretary can provide information on how to obtain the blood test. (Pullorum-typhoid testing, if required, is usually provided at the show.)

What Can Be Shown?

Any healthy duck of a standard breed can be exhibited at shows sanctioned by the American Poultry Association (APA). Most shows will also accept nonstandard breeds, although they cannot compete for awards against standard breeds. At some county and state fairs there are also classes for crossbreds.

These ducks are part of the Champion Row at the 1999 Northwest Winter Classic.

How Are Birds Entered?

To exhibit birds in a show, you need a copy of the premium list and an entry blank. For county, provincial, and state-fair premium lists and entry blanks, contact your local agriculture extension specialist. For other shows, check with established poultry breeders in your area and look over advertisements for shows in the *Poultry Press* or *Feather Fancier*. Normally, entry forms must be filled out and returned to the show secretary anywhere from 1 to 4 weeks prior to the show date, so you'll need to plan ahead. To avoid being disqualified, be sure to read and follow all instructions as outlined in the premium list.

Shows normally have separate classes for old drakes, old ducks, young drakes, and young ducks. Be careful to enter birds of the correct sex and age in the appropriate class. Some shows require birds to be tagged with numbered leg bands and these numbers are recorded on the entry blanks.

Selecting Show Ducks

Selecting show birds is a combination of art and science. To become proficient at selecting show ducks, it is necessary to learn the desired characteristics of your chosen breed. A good starting point is to study the section on breeds in this book and the *American Standard of Perfection*. Then, make a list of the most important three or four traits of the breed. For example, a list of the main characteristics for Call ducks would include:

1. Small size
2. Short bill
3. Large, round head
4. Wide, plump body

Novices often place too much emphasis on minor details that normally have little or no bearing on a bird's value as a show specimen. Examples of minor details that are often overemphasized include toes that are not perfectly straight (it's normal for the outside toe in particular on each foot to be somewhat curved) and minor variations in foot, bill, eye, or plumage color.

Keep in mind that type is more important than color in show birds. In the *American Standard of Perfection,* on a 100-point scale, color counts for 29 points in most white ducks and for 36 points in most color varieties. A bird with excellent type and size, but average color, will typically beat a bird with average type and size, but excellent color.

At many shows, a judge will appraise between 200 and 500 birds in a single day, meaning he or she can only spend a short time evaluating each bird. Winning ducks are strong in their breeds' distinctive characteristics and, as celebrated judge Henry Miller often said, their "quality meets the eye" as the judge walks down the aisle.

An excellent way to hone your evaluation skills is to prejudge your ducks and then compare against the judge's placement at the show. Most judges, once they are finished judging for the day, are willing to explain why they placed a class of birds as they did.

Preparing Ducks for Exhibition

To show to their best advantage, ducks must be clean, in good feather, carrying the correct amount of weight, and accustomed to being penned in a restricted enclosure. Birds that are dirty, in poor feather condition, or are over- or underweight will fare poorly under most judges, even if they are excellent specimens otherwise. Also, while appraising ducks, judges have little patience with birds that thrash wildly about or crouch in the corner of their cages.

Cage Training

For ducks that have always had the freedom to run around in a spacious pasture or yard, being locked up in a display cage can be a disconcerting experience. The first phase in training ducks for exhibition is to work calmly when near the birds and talk to them with a reassuring voice at feeding time. Several weeks prior to the show date, it's a good practice to coop birds that have never been exhibited in wire cages that are at least 24 inches wide x 24 inches deep x 27 inches tall for heavy breeds (cage size can be adjusted down for smaller breeds) for 2 or 3 days with food and drinking water. The cages should be located where you will walk by them occasionally throughout the day so the ducks become accustomed to people trafficking by on foot.

Conditioning

In shows, ducks are judged on a 100-point scale. The APA *Standard of Perfection* allots 10 of these points for condition and vigor. In good competition, the winners are clean and in excellent feather condition. Fortunately, ducks are neat birds and will clean themselves if given half a chance. Seldom do they need to be hand-washed prior to a show, except possibly to tidy the bill, feet, or a few soiled feathers with warm water and a soft brush (an old toothbrush is great for cleaning bills and feet) or sponge. Three of the worst enemies of show ducks are mud, overcrowding, and excessive exposure to the sunlight, which can cause the plumage to fade, dry out, and lose its sheen.

To encourage good condition, ducks should be kept in a clean environment, provided ample shade, and fed a diet that encourages lustrous feathering for 6 to 12 weeks prior to a show. (For more details on conditioning, see specific breed recommendations.)

Fitting Show Birds

Grooming birds for the show pen is an enjoyable and relaxing activity for many waterfowl exhibitors. Six to 8 weeks prior to an exhibition, show prospects can be examined for broken, stained, or faded feathers. If faulty feathers are plucked at this time, they should be regrown in time for the show. A day before the show, any small, off-colored body feathers can be removed, the toenails clipped and filed, the feet and bill washed with warm water, and a clean pan of bathing water provided in the pen for last-minute bathing. (As they are catching their ducks for a show, some people rub baby oil or some other substance onto the legs and feet, but great care must be taken to not get any on the feathers.)

Faking

Because poultry shows are partly beauty contests, exhibitors want their birds to look their best. However, artificially coloring any part of the bird and the cutting or removal of main wing and tail feathers are considered faking and can result in the disqualification of a bird.

On the other hand, the removal of off-colored body feathers (such as a few white feathers in a black duck) or the feeding of special feeds to enhance the color of feathers (for example, feeding yellow-pigmented ingredients to Pekins to increase the creamy yellow of their plumage) or bill are considered part of showing your ducks to their best advantage and are not faking.

Bird Ownership

Ducks must be owned by the person or farm under whose name the entry is made. 4-H, FFA, and some specialty club shows sometimes require a minimum length of ownership time prior to showing. However, most open-class shows have no minimum-length-of-ownership requirements.

Transporting Ducks to Shows

Ducks arrive at shows in many different kinds of containers, from cardboard boxes to fancy custom-designed carriers. The main concern in getting your birds to the show is that they have sufficient ventilation, clean bedding, and adequate space. For Indian Runners, I use crates a minimum of 28 inches tall to prevent them from bumping their heads, which can cause temporary neck kinking. An advantage of cardboard boxes is that they can be recycled when they are soiled; however, be sure to cut sufficient holes to provide good ventilation. Wooden crates are sturdier and can be decorated with the farm name or colors.

Homemade Show and Utility Cages

Handy wire cages — for show training, hospital isolation, quarantine, breaking up broody ducks, and other uses — can be made at home. Tools needed are wire-cutting pliers, cage clip applicator pliers, cage clips, 3-foot-tall welded wire fencing with 1 x 2-inch openings, and a flat metal file.

To make a single cage that is 3 feet long, 2¼ feet deep, and 3 feet tall (this size will accommodate even the tallest Runner and is easy to catch and remove birds from), purchase 15 feet of 3-foot-high welded wire and a 2-inch hasp.

Cut an 11-foot length of wire and bend it at right angles after 36 inches, another 30 inches, then another 36 inches to make the corners of the cage. The fourth corner is made by clipping together the two loose ends with half a dozen cage clips. For the top, cut a 30-inch length and clip it all the way around with cage clips.

On one of the 3-foot-long sides, measure up 6 inches from the bottom and cut a 16-inch-tall by 14-inch-wide opening in the center of the cage. (With the metal file, smooth any jagged wire ends so your arms will not be scratched as you take birds in and out of the cage.)

From the remaining original piece of wire, cut a door that is 20 inches tall and 16 inches wide. Position the door over the opening so that it overlaps one inch on each side and two inches top and bottom, and clip along one side. Use the 2-inch hasp to secure the door shut.

This cage can be set on a piece of plywood and bedded with wood shavings or straw, or welded wire with ½ x 1-inch openings can be clipped to the bottom for a wire floor.

Care at the Show

Some county and state fairs provide attendants who feed and water the birds throughout the duration of an exhibition. However, club shows do not normally provide this service and you are responsible for feeding and watering your entries. You should check ahead of time if this matter is not clarified in the premium list so you will not be caught without feed and containers if this service is not provided.

Showroom Etiquette

Most poultry shows are operated by volunteers who put in long hours and often significant amounts of their own money to make the shows happen. Whenever possible, thank the host club members and show workers for their efforts and ask if there is any way you can help. If everyone pitches in, shows become more enjoyable for all.

Unless judges initiate conversation, exhibitors should not communicate with them or in any way interfere during judging. Many shows allow spectators to observe the judging from a reasonable distance; however, observers should be quiet and not make loud comments about the quality or ownership of the birds.

Once judging is completed, most judges are willing to discuss why they placed the birds the way they did. This is not a time, however, to argue or attempt to change the placing of a bird. Remember, when you enter a bird in a show, you are agreeing to the appraisal of the hired judge. People who are good sports learn the most and receive the greatest enjoyment from shows and will congratulate the winners regardless of their personal feelings.

Interpreting Judging

One of the benefits of showing is having a disinterested person evaluate your birds. It is best not to take too seriously either the pros or cons of a single judge; even highly respected judges can disagree on the merits of a given bird. If breeders base decisions about which birds to breed from strictly on how they placed at a show, their breeding programs may be hampered.

I have watched many people dispose of valuable birds based on the evaluation of one judge. Once at a national waterfowl meet, I was standing with several junior exhibitors during the judging of a class of ducks that they were exhibiting. The judge made a number of derogatory comments about their birds, which prompted the young exhibitors to say that they were going to dispose of them. I suggested that their birds were of good quality and that they should show them at an upcoming show where a nationally known judge, more familiar with this breed, would be judging.

At the second show, I again stood with the young exhibitors as their birds were being evaluated. This time the judge specifically commented on how these birds were better than most he had seen of this relatively rare variety. Two well-known judges and two very different evaluations.

General Show Hints

Even out of the best breeding stock, only a portion of the offspring will be show specimens. When raising ducklings for show, plan to raise a minimum of two or three for each show bird desired. Most experienced breeders find that they must hatch anywhere from 15 to 100 ducklings to produce an elite show bird that is capable of taking top honors in stiff competition.

Although raising large numbers of ducklings generally increases your chances of producing an elite show bird, you should never raise more than you are capable of caring for properly. High-quality show birds are the result of good breeding, correct diet, and healthful environment.

Show Terms and Abbreviations

OF (old female): ducks over 1 year old

OM (old male): drakes over 1 year old

YF (young female): ducks under 1 year old

YM (young male): drakes under 1 year old

BB or **BOB**: Best of Breed

BOSB: Best Opposite Sex of Breed

BV or **BOV**: Best of Variety

BOSV: Best Opposite Sex of Variety

ASV: All Standard Varieties

Disq.: Disqualified

Young people can raise world-class ducks. Rebecca Simon shows off her champion Crested duck at the 1999 Western Waterfowl Expo.

APPENDIXES

APPENDIX A

MIXING DUCK RATIONS

For breeders who would like to mix their own duck rations, here are some examples
of balanced feeds for ducks at various stages of development.

Home-Mixed Starting Rations (0–2 Weeks)

INGREDIENT	No. 1 SMALL QUANTITY	No. 2 LARGE QUANTITY
Yellow cornmeal	11 cups	62 lbs.
Soybean meal (44% protein)	3½ cups	17 lbs.
Wheat bran	2 cups	2 lbs.
Meat and bone meal (50% protein)	½ cup	4 lbs.
Fish meal (60% protein)*	½ cup	2 lbs.
Alfalfa meal (17.5% protein)	½ cup	2 lbs.
Dried skim milk or calf manna	½ cup	3 lbs.
Brewer's dried yeast	1½ cups	7¼ lbs.
Dicalcium phosphate (18.5% phosphate)	1 tbsp.	½ lb.
Iodized salt	1 tsp.	¼ lb.
Cod liver oil**		
Totals	20 cups	100 lbs.
Chopped succulent greens	Free choice	Free choice
Sand or chick-sized granite grit	Free choice	Free choice
Chick-sized oyster shells or crushed dried eggshells	Free choice	Free choice

*If fish meal isn't available, use 4½ cups soybean meal and 10½ cups cornmeal in
formula #1, or 20 lbs. soybean meal and 61 lbs. cornmeal in formula #2.

**If birds do not receive direct sunlight, which enables them to synthesize vitamin D,
sufficient cod liver oil must be added to these rations to provide 500 International
Chick Units (ICU) of vitamin D_3 per pound of feed.

Home-Mixed Growing Rations (2–12 Weeks)

INGREDIENT	NO. 3 CORN BASE (LBS)	NO. 4 WHEAT BASE (LBS)
Cracked yellow corn*	75.0	—
Whole soft wheat	—	79.0
Milo (grain sorghum)	—	—
Soybean meal (50% protein)	9.0	5.0
Meat and bone meal (50% protein)	4.0	4.0
Alfalfa meal (17.5% protein)	3.0	3.0
Dried skim milk	3.0	3.0
Brewer's dried yeast	5.0	5.0
Dicalcium phosphate (18.5% phosphate)	0.10	0.10
Limestone flour or oyster shells	0.65	0.65
Iodized salt	0.25	0.25
Cod liver oil**		
Total (lbs.)	100.0	100.0
Sand or chick-sized granite grit	Free choice	Free choice
Chopped succulent greens (eliminate when pasture is available)	Free choice	Free choice

*Whole corn can be used after the birds are 4–6 weeks of age.
**See Home-Mixed Starting Rations (0–2 Weeks).

Home-Mixed Rations for Adult Ducks

INGREDIENT*	NO. 5 HOLDING (LBS)	NO. 6 HOLDING (LBS)	NO. 7 LAYER (LBS)	NO. 8 LAYER (LBS)
Whole milo or yellow corn	82.00	—	60.00	24.00
Whole soft wheat	—	86.00	9.00	48.00
Soybean meal (50% protein)	8.00	4.00	7.00	4.50
Meat and bone meal (50% protein)	—	—	4.00	4.00
Alfalfa meal (17.5% protein)	4.00	4.00	4.00	4.00
Dried skim milk	—	—	2.00	2.00
Brewer's dried yeast	5.00	5.00	7.00	7.00
Oyster shell	0.25	0.50	6.40	5.90
Dicalcium phosphate (18.5% protein)	0.50	0.25	0.30	0.35
Iodized salt	0.25	0.25	0.30	0.25
Cod liver oil**				
Totals (lbs.)	100.00	100.00	100.00	100.00

*The addition of 3–5 pounds of livestock-grade molasses to each 100 pounds of mixed feed reduces waste.

**If birds do not receive direct sunlight, which enables them to synthesize vitamin D, sufficient cod liver oil must be added to these rations to provide 400 International Chick Units (ICU) of vitamin D_3 per pound of feed.

Note: These rations are best suited for situations when birds have access to pasture.

Complete Rations for Ducklings (Pelleted)

INGREDIENT	NO. 9 CORN BASE STARTER (LBS/TON)	NO. 10 WHEAT BASE STARTER (LBS/TON)	NO. 11 CORN* BASE GROWER (LBS/TON)	NO. 12 WHEAT* BASE GROWER (LBS/TON)
Ground yellow corn	1385	—	1570	—
Ground soft wheat	—	1425	—	1607
Ground milo (grain sorghum)	—	—	—	—
Soybean meal solv. (50% protein)	430	360	295	270
Fish meal (60% protein)**	40	40	—	—
Meat and bone meal (50% protein)	80	80	80	40
DL-Methionine (98%)	2.00	2.75	1.50	2.00
Stabilized animal fat	25	40	—	20
Soybean oil	—	14	—	10
Dicalcium phosphate (18.5% phosphate)	6.25	5.25	8.00	17.00
Limestone flour (38% calcium)	6.00	7.00	20.00	9.00
Iodized salt	5.75	6.00	5.50	5.00
Vitamin:mineral premix	20	20	20	20
Totals (lbs.)	2000	2000	2000	2000

Vitamin:Mineral Premix

Vit. A (millions of IU/ton)	8.0	9.5	5.0	6.5
Vit. D_3 (millions of ICU/ton)	1.0	1.0	.8	.8
Vit. E (thousands of IU/ton)	5.0	15.0	2.0	12.0
Vit. K (g/ton)	2.0	2.0	2.0	2.0
Riboflavin (g/ton)	6.0	6.0	4.0	4.0
Vit. B_{12} (mg/ton)	8.0	8.0	4.0	4.0
Niacin (g/ton)	50.0	50.0	40.0	40.0
d-Calcium pantothenate (g/ton)	6.0	3.0	4.0	2.0
Choline chloride (g/ton)	300.0	—	200.0	—
Folic acid (g/ton)	.5	.5	.5	.5
Manganese sulfate (oz/ton)	9.6	9.6	9.6	9.6
Zinc oxide (80% Zinc, oz/ton)	3.2	3.0	3.2	3.0
Ground grain to make 20 lbs.	+	+	+	+
Totals (lbs.)	20.0	20.0	20.0	20.0

Calculated Analysis

Crude protein (%)	20.2	19.6	16.4	16.1
Lysine (%)	1.05	1.00	.76	.73
Methionine (%)	.46	.44	.37	.33
Metabolizable energy (kcal/lb)	1424	1366	1426	1362
Calorie:protein ratio	70	70	87	85
Crude fat (%)	4.7	4.7	3.5	3.3
Crude fiber (%)	2.3	2.4	2.3	2.4
Calcium (%)	.76	.78	.93	.63
Available phosphorus (%)	.41	.41	.36	.36
Vit. A (IU/lb)	4912	4750	3535	3250
Vit. D_3 (ICU/lb)	500	500	400	400
Riboflavin (mg/lb)	3.73	3.76	2.64	2.66
Total niacin (mg/lb)	41	51	31	43

*To convert into breeder-developer rations, substitute the following quantities of vitamins for those listed under the vitamin:mineral premix: 1 g of vit. K; 3 g riboflavin; 30 g niacin; and 100 g choline chloride in the corn and milo base rations.
**If fish meal is not available, use an additional 50 pounds soybean meal solvent (50% protein) and subtract 10 pounds of grain.

Complete Rations for Adult Ducks (Pelleted)

INGREDIENT	NO. 13 CORN BASE HOLDING (LBS/TON)	NO. 14 WHEAT BASE HOLDING (LBS/TON)	NO. 15 CORN BASE BREEDER (LBS/TON)	NO. 16 WHEAT BASE BREEDER (LBS/TON)
Ground yellow corn	1612	—	1415	—
Ground soft wheat	—	1667	—	1460
Ground milo (grain sorghum)	—	—	—	—
Soybean meal solv. (50% protein)	265	190	302	236
Meat and bone meal (50% protein)	—	—	80	80
Alfalfa meal (17.5% protein)	40	40	40	40
Stabilized animal fat	—	30	—	20
Soybean oil	—	5	—	15
DL-Methionine (98% methionine)	—	1.25	1.0	1.50
L-Lysine (50% lysine)	—	1	—	—
Dicalcium phosphate (18.5% phosphate)	28	27	12	10
Limestone flour (38% calcium)	30.00	13.75	125.00	112.50
Iodized salt	5	5	5	5
Vitamin:mineral premix	20	20	20	20
Totals (lbs.)	2000	2000	2000	2000

Vitamin:Mineral Premix

Vit. A (millions of IU/ton)	5.0	6.0	8.0	9.0
Vit. D$_3$ (millions of ICU/ton)	0.8	0.8	1.0	1.0
Vit. E (thousands of IU/ton)	—	10.0	10.0	20.0
Riboflavin (g/ton)	3.0	3.0	6.0	6.0
Vit. B$_{12}$ (mg/ton)	4.0	4.0	8.0	8.0
Niacin (g/ton)	30.0	30.0	50.0	50.0
d-Calcium pantothenate (g/ton)	2.0	1.0	6.0	3.0
Choline Chloride (g/ton)	100.0	—	300.0	—
Folic acid (g/ton)	—	—	0.5	0.5
Manganese sulfate (oz/ton)	9.6	9.6	9.6	9.6
Zinc oxide (80% Zinc, oz/ton)	3.2	3.0	3.2	3.0
Ground grain to make 20 lbs.	+	+	+	+
Totals (lbs.)	20.0	20.0	20.0	20.0

Calculated Analysis

Crude protein (%)	14.2	13.6	16.3	15.8
Lysine (%)	0.63	0.60	0.77	0.72
Methionine (%)	0.26	0.26	0.33	0.33
Methionine + cystine (%)	0.50	0.53	0.58	0.58
Metabolizable energy (kcal/lb)	1419	1359	1322	1278
Calorie:protein ratio	100	100	81	81
Crude fat (%)	3.3	3.4	3.2	3.6
Crude fiber (%)	2.6	2.8	2.6	2.7
Calcium (%)	0.95	0.64	2.99	2.75
Available phosphorus (%)	0.35	0.36	0.40	0.39
Vit. A (IU/ton)	5240	4678	6610	6178
Vit. D$_3$ (ICU/ton)	400	400	500	500
Riboflavin (mg/lb)	2.17	2.21	3.73	3.76
Total niacin (mg/lb)	25	38	35	46

KEY: IU = International Units   ICU = International Chick Units   g = gram
mg = milligram   Kcal = kilocalories

Appendix B

Symptoms of Vitamin and Mineral Deficiencies in Ducks

The symptoms for each deficiency below are listed in the order in which they usually manifest themselves.

Vitamins

Vitamin A. Retarded growth; general weakness; staggering gait; ruffled plumage; low resistance to infections and internal parasites; eye infection; lowered production and fertility; increased mortality. (Adult birds develop symptoms slower than ducklings.)
Sources: Fish-liver oils, yellow corn, alfalfa meal, fresh greens.

Vitamin D₃. Retarded growth; rickets; birds walk as little as possible, and when they do move, their gait is unsteady and stiff; bills become soft and rubbery and are easily bent; thin-shelled eggs; reduced egg production; bones of wings and legs are fragile and easily broken; hatchability is lowered.
Sources: Sunlight, fish-liver oils, synthetic sources.

Vitamin E. Unsteady gait; ducklings suddenly become prostrated, lying with legs stretched out behind, head retracted over the back; head weaves from side to side; reduced hatchability of eggs; high mortality in newly hatched ducklings; sterility in males and reproductive failure in hens.
Sources: Many feedstuffs both of plant and animal origin, particularly alfalfa meal, rice polish and bran, distiller's dried corn solubles, wheat middlings.

Vitamin K. Delayed clotting of blood; internal or external hemorrhaging, which may result in birds bleeding to death from even small wounds.
Sources: Fish meal, meat meal, alfalfa meal, fresh greens.

Thiamine. Loss of appetite; sluggishness; emaciation; head tremors; convulsions; head retracted over the back.
Sources: Grains and grain by-products.

Riboflavin. Diarrhea; retarded growth; curled-toe paralysis; drooping wings; birds fall back on their hocks; eggs hatch poorly.
Sources: Dried yeast, skim milk, whey, alfalfa meal, green feeds.

Niacin. Retarded growth; leg weakness; bowed legs; enlarged hocks; diarrhea; poor feather development.
Sources: Dried yeast and synthetic sources. (Most of the niacin in cereal grains is unavailable to poultry.)

Biotin. Bottoms of feet are rough and callused, with bleeding cracks; lesions develop in corners of mouth, spreading to area around the bill; eyelids eventually swell and stick shut; slipped tendon (see also choline and manganese deficiencies); eggs hatch poorly.

Sources: Most feedstuffs, but especially dried yeast, whey, meat and bone meal, skim milk, alfalfa meal, soybean meal, green feeds. (The biotin in wheat and barley is mostly unavailable to poultry. If raw eggs are fed to animals, avidin — a protein in the egg white — binds biotin, making it unavailable.)

Pantothenic Acid. Retarded growth; viscous discharge causes eyelids to become granular and stick together; rough-looking feathers; scabs in corners of mouth and around vent; bottoms of feet rough and callused, but lesions are seldom as severe as in a biotin deficiency; drop in egg production; reduced hatchability of eggs; poor livability of newly hatched ducklings.

Sources: All major feedstuffs, particularly brewer's and torula yeasts, whey, skim and buttermilk, fish solubles, wheat bran, alfalfa meal.

Choline. Retarded growth; slipped tendons (see also biotin and manganese deficiencies).

Sources: Most feedstuffs but especially fish meal, meat meal, soybean meal, cottonseed meal, wheat germ meal. (Evidence indicates that choline is synthesized by mature birds in quantities adequate for egg production.)

Vitamin B_6. Poor appetites; extremely slow growth; nervousness; convulsions; jerky head movements; ducklings run about aimlessly, sometimes rolling over on their backs and rapidly paddling their feet; increased mortality.

Sources: Grains and seeds.

Folacin. Retarded growth; poor feathering; colored feathers show a band of faded color; reduced egg production; decline in hatchability; occasionally slipped tendon.

Sources: Green feeds, fish meal, meat meal.

Vitamin B_{12}. Poor hatchability; high mortality in newly hatched ducklings; retarded growth; poor feathering; degenerated gizzards; occasionally slipped tendon.

Sources: Animal products and synthetic sources.

MINERALS

Calcium and Phosphorus. Rickets; retarded growth; increased mortality; in rare cases, thin-shelled eggs.

Sources: Calcium — oyster shells and limestone. Phosphorus — dicalcium phosphate, soft rock phosphate, meat and bone meal, fish meal. (Most all feedstuffs have varying amounts of calcium and phosphorus, but typically only about one-third of the phosphorus in plant products is available to birds.)

Magnesium. Ducklings go into brief convulsions and then lapse into a coma from which they usually recover if they are not swimming; rapid decline in egg production.

    Sources: Most feedstuffs, especially limestone, meat and bone meals, grain brans. (Raising either the calcium or phosphorus content of feed magnifies a deficiency of this mineral.)

Manganese. Slipped tendon in one or both legs; retarded growth; weak egg shells; reduced egg production and hatchability. Slipped tendon (also known as perosis) is first evidenced by the swelling and flattening of the hock joint, followed by the Achilles tendon slipping from its condyles (groove), causing the lower leg to project out to the side of the body at a severe angle. (Perosis can also be the result of biotin or choline deficiencies.)

    Sources: Most feedstuffs, especially manganese sulfate, rice bran, limestone, oyster shell, wheat middlings, and bran.

Chloride. Extremely slow growth; high mortality; unnatural nervousness.

    Sources: Most feedstuffs, especially animal products, beet molasses, alfalfa meal.

Copper and Iron. Anemia.

    Sources: Most feedstuffs, including fresh greens.

Iodine. Goiter (enlargement of the thyroid gland); decreased hatchability of eggs.

    Sources: Iodine. Iodized salt contains such small quantities of iodine that it cannot be relied upon to provide sufficient iodine.

Potassium. Rare, but when it occurs, results in retarded growth and high mortality.

    Sources: Most feedstuffs.

Sodium. Poor growth; cannibalism; decreased egg production.

    Sources: Most feedstuffs, but especially salt and animal products.

Zinc. Retarded growth; moderately to severely frayed feathers; enlarged hock joints; slipped tendon.

    Sources: Zinc oxide and most feedstuffs, especially animal products.

Summarized from *Nutrient Requirements of Poultry*, 7th revised edition (1977), pages 11–20, with the permission of the National Academy of Sciences, Washington, D.C.

Note: While work on vitamin and mineral deficiencies in poultry has been limited largely to chickens and turkeys, I have observed many of these symptoms in ducks.

Appendix C

Predators

One of the most frustrating experiences for the owner of ducks is to lose valuable birds and eggs to predators. It is a good idea to talk with poultry keepers in your area to learn what kind of problems they have encountered and what precautions are necessary to minimize losses.

Of domestic animals, dogs — and to a lesser extent, cats — are notorious for the damage they can inflict on poultry flocks. The first time one of these pets shows an interest in your ducks, they must be disciplined *immediately* if future troubles are to be avoided.

Some of the wild creatures that cause duck raisers grief are rats, ground squirrels, weasels, mink, skunks, raccoons, opossum, foxes, coyotes, turtles, snakes, crows, ravens, magpies, jays, gulls, hawks, falcons, and owls. Because these animals are essential in maintaining the delicate balance of nature, we must not attempt to eliminate them, but should respect them and give poultry sufficient protection so that hungry predators will not be able to dine at the expense of our birds.

Security Measures

Ducks roost on the ground, making them most vulnerable at night. Because many predators have nocturnal habits, a small building or fenced yard where ducks can be locked in after dark is a *must* in most localities. If you raise ducks without penning them up nightly, it is almost 100-percent certain that your flock will be ravaged sooner or later.

A sturdy woven-wire fence at least 4 feet high goes a long way in keeping ducks safe while they sleep. In areas with determined hunters such as raccoons and opossum, electric fencing can be used in combination with woven-wire fences. To be the most effective, two strands of electric wire should be used. One strand should run around the outside of the woven wire 4 inches above the ground, while the second strand should be installed a couple of inches above the top of the fence.

If weasels, mink, owls, and cats (wild or domestic) are prevalent in your area, pen your ducks at night in a shelter having a covered top as well as sturdy sides. Some predators will dig under fences or dirt floors of bird houses. Burying the bottom 6 to 12 inches of fences that encircle yards or covering the floors of duckhouses with wire netting will keep out excavators.

Because setting hens and ducklings are especially vulnerable to predation, extra care must be taken to provide them with secure quarters. Setting hens should be encouraged to nest in shelters that can be closed at night. When hens do nest in the open, panels 4 feet high can be set up around them to provide protection.

Identifying the Culprit

The following is a brief guide to help you recognize the work of some of the more common predators, with suggestions on how to stop their thievery.

Dogs. *Telltale signs:* A number of ducks, sometimes the entire flock, badly maimed. Check for large holes under or through fences, and clumps of dog hair caught on wire. *Stopping losses:* Strong fences at least 4 feet high.

Cats (wild or tame). *Telltale signs:* Crushed eggs held together by shell membranes. Birds disappear without a trace, or only a few feathers or clumps of down are found in a secluded spot where the animal fed. *Stopping losses:* Keep ducklings in wire-covered runs until they are 2 to 4 weeks old. When a pet cat is seen stalking your birds, let the tabby know immediately that the ducks are off limits. Throwing a rolled up newspaper at the offending animal will usually get the message across. If you have problems with stray cats, a live trap may be needed.

Foxes. *Telltale signs:* Foxes are fastidious hunters and normally leave little evidence of their visits. Usually they kill just one duck at a time and take the bird with them or partially bury it nearby. (Foxes have been known to go on rampages, killing 30 or more birds at a time and scattering carcasses over a quarter-mile area.) You might be able to find a small hole under or through fences, or a poorly fitted gate or door pushed ajar. *Stopping losses:* Tight fencing at least 4 feet high with two strands of barbed or electric wire and close-fitting gates. Foxes will squeeze or dig under fences that are not flush with the ground.

Raccoons. *Telltale signs:* End of eggs bitten off, or crops (esophagus) eaten out of dead birds; possibly heads missing. Usually returns every fourth or fifth night. *Stopping losses:* Fences 4 feet high with two strands of electric wire, or lock birds at night in shelter with covered top and sides.

Skunks. *Telltale signs:* Destroyed nest with crushed shells mixed with nest debris. *Stopping losses:* Gather eggs daily. Encourage setting hens to nest in shelters that can be locked up at night, or set up panels around hens that are nesting out in open areas.

Opossum. *Telltale signs:* Smashed eggs and birds that are badly mauled. *Stopping losses:* At night, lock birds in a shelter with covered top and sides.

Mink and Weasels. *Telltale signs:* Young ducklings disappear or larger birds are killed, evidently for amusement, with small teeth marks on head and neck. *Stopping losses:* At night, lock the birds in a shelter that is covered with ½-inch wire hardware cloth. As incredible as it seems, mink and weasels can pass through holes as small as 1 inch in diameter.

Snapping Turtles and Large Fish. *Telltale signs:* Ducklings disappear mysteriously while swimming. *Stopping losses:* If your ducks frequent bodies of water that host turtles or large fish such as northern pike and large-mouth bass, keep ducklings away from the water until they are 2 to 4 weeks old.

Snakes. *Telltale signs:* Nest appears untouched, but some or all eggs are missing. *Stopping losses:* Encourage setting hens to lay in covered nest boxes.

Rats. *Telltale signs:* Eggs or dead ducklings pulled into underground tunnels. *Stopping losses:* Until ducklings are 3 to 6 weeks old, at night put them in a pen that has sides, top, and floor covered with ½-inch wire mesh. Rat populations should be kept under control with cats or traps.

Crows, Jays, Magpies, and Gulls. *Telltale signs:* Punctured eggs or shells scattered around base of an elevated perch such as a fencepost or tree stump. These birds also occasionally steal newly hatched ducklings. *Stopping losses:* Provide covered nest boxes and gather eggs several times a day. Keep ducklings in wire-covered runs until they are at least 2 weeks old.

Hawks. *Telltale signs:* Ducklings disappear during daylight without a trace or only a few scattered feathers or clumps of down. Hawks are probably falsely accused of stealing poultry more often than any other predator. Because the large, soaring hawks of the *Buteo* genus are so visible, people mistakenly assume these winged hunters are the cause of every missing barnyard fowl. Actually, the hawks we need to worry about are the ones that are seldom seen by the casual observer: the accipiter family, which includes the goshawk, Cooper's hawk, and sharp-shinned hawk. The accipiters are secretive, stick to trees, and hunt from low altitudes. They can be identified by their short, round wings and long tails. If you do occasionally lose ducklings to hawks, your frustration may be reduced if you remember that the average hawk eats 200 to 300 rodents yearly. *Stopping losses:* Keep ducklings in wire-covered runs until they are 2 to 4 weeks old.

Owls. *Telltale signs:* One or more ducks killed nightly with head and neck eaten. *Stopping losses:* At night, lock ducklings and adult ducks in a shelter covered with 1-inch x 1-inch wire netting. Some of the smaller species (such as screech owls) can squeeze through 2-inch x 2-inch wire mesh.

APPENDIX D

DUCK RECIPES

There are many ways duck can be fixed to taste great. Unfortunately, duck cookery is ignored or passed over lightly in most cookbooks. So with a lot of help from my wife, Millie, I'd like to share a few of our favorite recipes.

Fried Duckling
Serves 4 to 6

    1 duckling, cut up into pieces
    ½ cup flour
    1 tsp. salt
    ⅛ tsp. pepper

Dust pieces of duckling with flour. Season with salt and pepper. Fry in lightly greased skillet over medium-low heat for 1½ hours. Turn pieces as necessary to brown evenly. Remove fat from skillet as it fries out of duck. Use drippings for gravy.

Roast Duckling
Serves 4 to 6

    1 whole 4½- to 5-lb. duckling
    1 tsp. salt
    ⅛ tsp. pepper
    3–4 cups homemade or 1 small box store-bought stuffing

Rub inside of duck cavity with salt and pepper. The duck may be roasted with or without stuffing. (In-bird stuffing often does not receive enough heat to cook thoroughly. To avoid the risk of illness from undercooking, make certain stuffing is cooked completely in bird, or prepare stuffing as a side dish.) Fasten opening with skewers and lace closed with strong thread or thin string.

Place duckling uncovered on rack of roasting pan. Roast in preheated oven at 325°F for 2½ to 3 hours, until skin is crisp and brown and flesh is tender.

Steam-Fried Duck Eggs

Put eggs into medium-hot, lightly greased skillet. Pour a small amount of hot water around eggs and place lid on skillet. Steam-fry for several minutes until egg white is set. Serve immediately.

Duck Stew

Serves 6 to 8

2 cloves garlic, minced
2 tbsp. butter or margarine
1–2 cups leftover duck meat, in chunks
3 medium-sized potatoes, diced
4 carrots, sliced in ½-inch pieces
1 large onion, wedged
1 cup shredded cabbage

2 cups tomatoes
1–2 quarts vegetable or meat stock
Salt, pepper, paprika, and herbs, to taste
1 cup peas, green beans, or lima beans
2–3 celery stalks, cut in chunks
1 cup corn

In a large kettle, sauté garlic in butter or margarine. Add the duck meat, potatoes, carrots, onion, cabbage, and tomatoes. Cover with vegetable or meat stock. Season with salt, pepper, paprika, and desired herbs.

Cook over medium-low heat until vegetables are crispy-tender. Add water if necessary to keep vegetables covered.

Add the beans, celery, and corn. Heat to boiling point (or longer if fresh or frozen beans are used). If you prefer a thicker stew, add a paste of flour and water. Serve with fresh, warm homemade bread.

Whole-Wheat Angel Food Cake

1½ cups duck egg whites (8–10 eggs)
1½ tsp. cream of tartar
¼ tsp. salt
1 tsp. vanilla
½ tsp. almond flavoring
¾ cup packed brown sugar
1 cup sifted whole wheat flour

In a large bowl, beat together (until stiff but glossy) the egg whites, cream of tartar, salt, vanilla, and almond flavoring.

Add the brown sugar to beaten whites, ¼ cup at a time, beating well after each addition. Fold in the flour with a large spoon, sifting a little over the top, folding in lightly with a down-up-over motion.

When well blended, pour into an ungreased 10-inch angel food cake pan. Bake at 375°F for 45 to 60 minutes; touch the top gently to see if cake is done. Invert pan and cool cake thoroughly. Remove from pan.

Option: Substitute ¼ cup carob or cocoa powder for an equal amount of flour and omit almond flavoring.

APPENDIX E

USING FEATHERS AND DOWN

Those soft feathers and down that were carefully saved at butchering time can be used to make a variety of handy items. While all ducks produce good-quality plumage, the down of Muscovies is less desirable than that of the breeds derived from Mallards. Large breeds such as Aylesbury, Pekin, and Rouen will produce 2½ to 3½ ounces of down and small feathers per bird.

People frequently ask about plucking live ducks for down. While ducks usually survive such treatment, this practice places a great deal of stress on the birds and reduces their productivity. If ducks are live-plucked, taking the following precautions will reduce the negative effects it has on the birds: 1) Pick the birds only once a year during the late spring or early summer after they have begun their natural molt. At this time, their feathers will be loosened, and the plumes can be removed with less pain to the birds. 2) Pull out only small pinches of feathers and down at a time to keep from tearing the bird's skin. 3) Take feathers only from the underside of the duck. 4) Remove a maximum of 50 percent of the feathers from the plucked area, and do not leave any bare patches. 5) Do not let the birds swim for 2 or 3 weeks, or until they have grown new feathers.

When making down pillows, a tightly woven cloth, such as down-proof ticking, and double-stitched seams are essential to keep down and feathers from working their way out. Small, soft body feathers need to be mixed with down to make pillows more resilient. An excellent ratio is 75 percent down and 25 percent feathers, although good pillows can be made with a mixture of half down and half feathers.

A down-filled coverlet is not only delightful on a winter night, but is also efficient. The tops and bottoms of comforters, quilts, or sleeping bags must be lined with down-proof material. To keep the down and feathers distributed evenly, make channels 5 to 6 inches wide. Leave one end of the channels open for stuffing.

If the large plumes of the wings, tail, and body have been discarded at picking time, feathers and down can be used in the same ratio in which they come off the bird. As each channel is filled, sew the opening shut by hand, then finish the edges with binding.

For down clothing, as with comforters, you will need to make channels in order to keep the warmth evenly distributed. For parkas, vests, and other items, make the channels narrower, approximately 2 to 3 inches wide. Remember to use down-proof lining and to double those seams so that none of the down escapes. To keep clothing as lightweight as possible, use 75 to 90 percent down, mixed with small quantities of body feathers.

APPENDIX F

DUCK BREEDERS AND HATCHERY GUIDE

This guide is provided for your convenience, and not as an endorsement of individual breeders.

Cackle Hatchery, P.O. Box 529, Lebanon, MO 65536; (417) 532-4581. Khaki Campbell, Mallard, Pekin, Rouen, Fawn & White Runner, Swedish.

Clearview Hatchery, P.O. Box 399, Gratz, PA 17030; (717) 365-3234. Calls (Gray, Snowy, White), Khaki Campbell, East Indie, Mallard, Pekin, Rouen, Fawn & White Runner, Swedish.

Country Hatchery, Route 1, Box 174, Wewoka, OK 74884; (405) 257-8231. Pekin, Rouen.

Dunlap Hatchery, Box 507, Caldwell, ID 83606-0507; (208) 459-9088. Cayuga, Pekin, Rouen, Swedish.

Eagle Nest, P.O. Box 504, Oceola, OH 44860; (419) 562-1993. Khaki Campbell, Mallard, Pekin, Rouen, Runners, Swedish.

Grain Belt Hatchery, Box 125, Windsor, MO 65360; (660) 647-2711. Khaki Campbell, Mallard, Pekin, Rouen, Runners.

Harder's Hatchery, 624 N. Cow Creek Road, Ritzville, WA 99169; (509) 659-1423. Khaki Campbell, Cayuga, Orpington, Pekin, Rouen, Runners, Swedish.

Hoffman Hatchery, Box 128, Gratz, PA 17030; (717) 365-3694. Khaki Campbell, Cayuga, Mallard, Muscovy, Pekin, Rouen, Runners, Swedish, Welsh Harlequin.

Holderread's Waterfowl Preservation Center & Farm, P.O. Box 492, Corvallis, OR 97339; (541) 929-5338. Ancona, Silver Appleyard, Australian Spotted, Aylesbury, Khaki Campbell, Cayuga, Dutch Hookbill, East Indie, Magpie (Black, Blue), Muscovy (various colors), Rouen (Gray, Blue Fawn, Pastel), Runner (15 colors), Saxony, Silky Ducks, Welsh Harlequin.

Hoover's Hatchery, P.O. Box 200, Rudd, IA 50471; (800) 247-7014 or (515) 395-2730. Mallard, Pekin, Rouen.

Ideal Poultry Breeding Farm, P.O. Box 591, Cameron, TX 76520-0591; (254) 697-6677. Khaki Campbell, Cayuga, Crested, Mallard, Orpington, Pekin, Rouen, Fawn & White Runner, Swedish.

Inman Hatcheries, P.O. Box 616, Aberdeen, SD 57402-0616; (800) 843-1962. Khaki Campbell, Cayuga, Crested, Mallard, Orpington, Pekin, Rouen, Fawn & White Runner, Swedish.

Kruse Hatchery, 1011 CR W 14, Ft. Atkinson, IA 52114; (319) 534-7396. Pekin, Rouen.

Marti's Hatchery, P.O. Box 27, Windsor, MO 65360; (660) 647-3156. Khaki Campbell, Mallard, Pekin, Rouen, Runners.

Metzer Farms, 26000 Old Stage, Gonzales, CA 93926; (800) 424-7755. Khaki Campbell, Cayuga, Crested, Hybrids, Mallard, Orpington, Pekin (various strains), Rouen, Runner (Black, Blue, Chocolate, Fawn & White), Swedish.

Mt. Healthy Hatcheries, Inc., 9839 Winton Rd., Mt. Healthy, OH 45231; (800) 451-5603. Mallard, Pekin.

Murray McMurray Hatchery, P.O. Box 458, Webster City, IA 50595; (800) 456-3280 or (515) 832-3280. Khaki Campbell, Cayuga, Crested, Mallard, Orpington, Pekin, Rouen, Fawn & White Runner, Swedish.

Phinney Hatchery, 1331 Dell Ave., Walla Walla, WA 99362; (509) 525-2602. Khaki Campbell, Cayuga, Orpington, Pekin, Rouen, Swedish.

Privette's Hatchery, P.O. Box 176, Portales, NM 88130; (800) 634-4390 or (505) 356-6425. Khaki Campbell, Cayuga, Crested, Mallard, Orpington, Pekin, Rouen, Runners, Swedish.

Ridgeway Hatcheries, P.O. Box 306, LaRue, OH 43332; (800) 323-3825 or (740) 499-2163. Khaki Campbell, Mallard, Muscovy, Pekin, Rouen, Runners, Swedish.

Sand Hill Preservation Center, 1878 230th St., Calamus, IA, 52729; (319) 246-2299. Ancona, Silver Appleyard, Campbell (Dark, Golden, Khaki, Pied), Golden Cascade, Cayuga (Black, Blue), Crested, Magpie, Muscovy, Orpington, Pekin, Runners, Saxony, Swedish, Welsh Harlequin.

Shank's Hatchery, P.O. Box 429, Hubbard, OR 97032; (503) 981-7801. Khaki Campbell, Cayuga, Mallard, Pekin, Rouen, Swedish.

Stromberg's, P.O. Box 400, Pine River, MN 56474; (800) 720-1134. Silver Appleyard, Aylesbury, East Indie, Call (Gray, Snowy, White), Khaki Campbell, Cayuga, Crested, Mallard, Muscovy, Orpington, Rouen, Fawn & White Runner, Swedish (Black, Blue, Silver), Welsh Harlequin.

Sun Ray Chicks, P.O. Box 300, Hazelton, IA 50641; (319) 636-2244. Pekin, Rouen.

Townline Poultry Farm, P.O. Box 108, Zeeland, MI 49464; (616) 772-6514. Khaki Campbell, Pekin, Rouen.

Urch/Turnland Poultry, 2142 NW 47 Ave., Owatonna, MN 55060-1071; (507) 451-6782. Silver Appleyard, Aylesbury, Khaki Campbell, Cayuga, Crested, Magpie (Black, Blue), Mallard (Snowy), Muscovy (Black, Blue, Chocolate), Orpington, Pekin, Rouen, Runner (Black, Fawn & White, Mallard, Penciled, White), Swedish.

Welp's Hatchery, P.O. Box 77, Bancroft, IA 50517; (800) 458-4473. Khaki Campbell, Cayuga, Crested, Mallard, Orpington, Pekin, Rouen, Fawn & White Runner, Swedish.

APPENDIX G

SOURCES OF SUPPLIES AND EQUIPMENT

Sources in this appendix are given for your convenience, not as an endorsement.

A.B. Incubators Ltd., P.O. Box 215, Moline, IL 61265; (309) 793-4273, fax (309) 793-4286. Small incubators.

Allen Publishing LLC, Animal Products Division, P.O. Box 171227, Salt Lake City, UT 84117. Tek-Trol disinfectant and Vionate vitamins and minerals.

BF Products Inc., P.O. Box 61866, Harrisburg, PA 17106-1866; (717) 238-7715, fax (717) 238-7725 or (800) 255-8397; E-mail: BFPROD@paonline.com. Fencings and nettings.

Bowles Poultry Supply, 312 O'Connor Road, Lucasville, OH 45648; (740) 372-3973. Bands, medications, books, incubators, and other supplies.

Brinsea Products Inc., 3670 S. Hopkins Avenue, Titusville, FL 32780; (407) 267-7009 or (888) 667-7009, fax (407) 267-6090; Web site: www.brinsea.co.uk. Small incubators.

Brower Manufacturing Co., P.O. Box 2000, Houghton, IA 52631; (319) 469-4141. Brooders, waterers, feeders.

Clausing Company, Nocatee, FL 34268; (941) 993-2542. Bands, medications, books, incubators, and other supplies.

Collapsible Wire Products, 5120 N. 126 Street, Butler, WI 53007; (262) 781-6125. Wire exhibition cages.

Cutler's Supply, 3805 Washington Road, Carsonville, MI 48419; (810) 657-9450. Bands, medications, books, incubators, and other supplies.

First State Veterinary Supply, P.O. Box 190, Parsonburg, MD 21849; (800) 950-8387. Medications.

Gey Band and Tag Co., Inc., P.O. Box 363, Norristown, PA 19404; (610) 277-3280. Bands and tags.

G.Q.F. Manufacturing Company, 2343 Louisville Road, P.O. Box 1552, Savannah, GA 31498; (912) 236-0651, fax (912) 234-9978. Wire pens, incubators, brooders, supplies.

The Humidaire Incubator Company, P.O. Box 9, New Madison, OH 45346; (937) 996-3001 or (800) 410-6925, fax (937) 996-3633; E-mail: hatch@bright.net. Redwood cabinet incubators.

J.A. Cissel Mfg. Co., P.O. Box 2025, Lakewood, NJ 08701; (732) 901-0300 or (800) 631-2234, fax (732) 901-1166; E-mail: jacissel@Compuserve.com. Netting, grid flooring, rigid fencing, catch nets, and other supplies.

Jeffers Vet Supply, P.O. Box 100, Dothan, AL 36302; (800) 533-3377. Medications and general supplies.

Kalglo Electronics Co., Inc., 5911 Colony Drive, Bethlehem, PA 18017-9348; (610) 837-0700 or (888) 452-5456, fax (610) 837-7978; E-mail: kalglo@kalglo.com; Web site: www.kalglo.com. Infrared brooder heaters.

Keipper Cooping Company, P.O. Box 249, Big Bend, WI 53103; (262) 662-2290. Wire exhibition cages.

Kuhl Corporation, P.O. Box 26, 39 Kuhl Road, Flemington, NJ 08822; (908) 782-5696. Incubators, feeders, waterers.

Lyon Electric Company, 1690 Brandywine Avenue, Chula Vista, CA 91911; (619) 216-3400. Incubators, brooders.

Max-Flex, U.S. Route 219, Lindside, WV 24951; (304) 753-4387 or (800) 356-5458. Electroplastic fencing.

Mike Consumer Products, P.O. Box 4000, Blue Mountain, AL 36204; (205) 237-9461, fax (205) 237-8816. Netting for cage tops.

Nasco Farm and Ranch, 901 Janesville Avenue, Fort Atkinson, WI 53538-0901; (414) 563-2446 or (800) 558-9595. General farm supplies.

National Band & Tag, 721 York Street, Box 72430, Newport, KY 41072; (859) 261-2035. Bands and tags.

NatureForm Hatchery Systems, 1310 Tradeport Drive, Jacksonville, FL 32218; (904) 741-3030 or (800) 282-6252, fax (904) 741-4209; E-mail: natureform@aol.com. New and used incubators.

Patterson Poultry Supplies, 210 Meadowbrook Lane, Martinsville, VA 24112; (540) 638-2297. Bands, medications, books, incubators, and other supplies.

Petersine Incubator Co., P.O. Box 308, 300 North Bridge St., Gettysburg, OH 45328; (937) 447-2151 or (888) 255-0067, fax (937) 447-7171. Medium to large incubators.

PMI Feeds, Inc., 1401 South Hanley Road, St. Louis, MO 63144; (800) 227-8941. Mazuri waterfowl feeds for all ages.

Premier Fence Supply, 2031 300th Street, Washington, IA 52353; (319) 653-7622 or (800) 282-6631. Electroplastic fencing.

Seven Oaks Game Farm and Supply, 7823 Masonboro Sound Road, Wilmington, NC 28409-2672; (910) 791-5352, fax (910) 452-7807; E-mail: gamefarm@isaac.net. Bands, medications, books, incubators, pond liners, and other supplies.

Shenandoah Mfg. Co., Inc., 107 Virginia Avenue, Harrisonburg, VA 28802; (800) 476-7436, fax (800) 434-3068; Web site: www.poultry-equipment.com. Brooders, nests, feeders, waterers, cremators, and other supplies.

Smith Poultry & Game Bird Supply, 14000 West 215 Street, Bucyrus, KS 66013; (913) 879-2587. Bands, medications, books, incubators, and other supplies.

Stromberg's Chicks and Game Birds Unlimited, P.O. Box 400, Pine River, MN 56474; (218) 587-2222. Bands, medications, books, incubators, Dux-Wax, and other supplies.

Tomahawk Live Trap Co., P.O. Box 323, Tomahawk, WI 54487; (715) 453-3550. Humane predator traps.

Waterford Corporation, 404 North Link Lane, Fort Collins, CO 80524; (970) 482-0911 or (800) 525-4952. Electroplastic fencing.

Appendix H

Organizations and Publications

General Organizations

American Poultry Association, 133 Millville Street, Mendon, MA 01756; Web site: www.ampltya.com

American Bantam Association, P.O. Box 127, Augusta, NJ 07822

Heritage Breed Organizations

American Livestock Breeds Conservancy, P.O. Box 477, Pittsboro, NC 27312; (919) 542-5704

Rare Breeds Canada, c/o Dr. Tom Hutchinson, Trent University, Box 4800, Peterboro, Ontario K9J 7B8 Canada; (705) 748-1634

Rare Breeds Survival Trust, Attn: Peter King, National Agriculture Centre, Stoneleigh Park, Warwickshire CV8 2LC England; 01144-1203-696-551

Society for the Preservation of Poultry Antiquities, c/o Glenn Drowns, 1878 230th Street, Calamus, IA 52729; (319) 246-2299

Publications

Fancy Fowl, Barn Acre House, Saxtead Green, Nr. Woodbridge, Suffolk, England IP 13 9QJ. Monthly color magazine for poultry and waterfowl hobbyists.

Feather Fancier, 1137 Telfer Side Road, Sarnia, Ontario, Canada N7T 7H2; (519) 542-4963. Monthly newspaper and source for clubs, shows, and suppliers of birds and equipment.

Game Bird and Conservationists' Gazette, P.O. Box 171227, Salt Lake City, UT 84117. Bimonthly magazine and source for suppliers of wild ducks and equipment.

Poultry Press, P.O. Box 542, Connersville, IN 47331-0543; (765) 827-0932. Monthly newspaper and source for clubs, shows, and suppliers of birds and equipment.

GLOSSARY

Air cell. A pocket of air that develops between the two shell membranes in the large end of eggs shortly after they are laid.

Bean. The hard nail at the tip of a duck's bill.

Breed. A subdivision of the duck family whose members possess similar body shape and size, and the ability to pass these characteristics on to their offspring.

Breed true. When offspring resemble their parents.

Breeder ration. A feed used for the production of hatching eggs.

Broiler. Quick-grown poultry used for meat at an early age.

Buffled. Having the fluffy head feathering desired in Call ducks.

Candling. The illumination of an egg's contents with a bright light.

Caruncles. The bumpy flesh on the face of Muscovies.

Class. The duck breeds of similar size that are grouped together in the *American Standard of Perfection*.

Cloaca. The inner cavity behind the vent where the urinary, intestinal, and reproductive canals open in fowl.

Concentrated feed. Feeds that are high in protein, carbohydrates, fats, vitamins, and minerals, and low in fiber.

Crest. The elongated feathers on the head of some breeds of ducks.

Down. The fur-like covering of newly hatched ducklings.

Drake. A male duck.

Drakelet. A young male duck.

Duck. In general, any member of the Anatidae family; it is often used specifically in reference to females of the duck family.

Ducklet. A young female duck.

Ducklings. Young ducks up until their feathers have completely replaced their baby down.

Eclipse molt. A 3- to 4-month period each year, after the breeding season, when the bright plumage of colored adult drakes is replaced with subdued colors similar to those of females.

Egg tooth. Small, horny protuberance attached to the bean of the bills of newly hatching birds that is used to help break the shell at hatching time. It falls off several days after hatching.

Fault. Any characteristic of a purebred duck that falls short of the written standard.

Feed conversion. The ability of birds to convert feed into body growth or eggs. To calculate feed conversion ratios, divide pounds of feed consumed by pounds of body weight or eggs.

Fertile. Sexually mature ducks that are capable of reproducing; or an egg that contains a viable embryo.

Fertility. In reference to eggs, the capability of producing an embryo. Fertility is expressed as a percentage that equals the total number of eggs set minus those that are infertile, divided by the total number set, times 100.

Flights. The large feathers of the wing, including the primaries and the secondaries.

Flock mating. A group of breeding ducks that are penned together and consist of more than one male and two females.

Green ducklings. Young ducks that are managed for fast growth and then slaughtered at 7 to 8 weeks of age.

Growing ration. A feed that is formulated to stimulate fast growth in ducklings over 2 weeks old.

Hatchability. The ability of eggs to hatch. Hatchability can be expressed as 1) a percentage of the fertile eggs set (total number of ducklings hatched divided by the number of fertile eggs set, times 100) or 2) a percentage of all eggs set (total number of ducklings hatched divided by the total number of eggs set, times 100).

Hybrid. The offspring of a planned breed cross; or the offspring of parents of different species.

Incubation period. The number of days it takes eggs to hatch once they are warmed to incubation temperature.

Infertile. Eggs that do not contain an embryo; or mature animals that are incapable of reproduction.

Keel. The pendulous fold of skin hanging from the underbody of some ducks.

Laying ration. Feed that is formulated to stimulate high egg production.

Linebred. A planned form of inbreeding that conserves the genetic material of specific ancestors.

Maintenance ration. A feed used for adult ducks that are not in production.

Molt. The natural replacement of old feathers with new ones.

Nuptial plumage. In colored varieties of ducks, the bright breeding plumage of males exhibited during fall, winter, and spring.

Oil gland. Also known as uropygial gland. Located at the base of the tail, it produces water-repellent oil for the feathers.

Old drake or old duck. A drake or duck more than one year old.

Oviduct. The tube that transports the egg from the ovary to the cloaca.

Pair. A breeding unit consisting of a male and female.

Pinfeathers. New feathers that are just emerging from the skin.

Pip. The first visible break the duckling makes in the eggshell.

Primaries. The ten large, outermost feathers of each wing.

Production-bred. Ducks that have been selected for top meat and/or egg production.

Pullet. A young fowl in her first year of egg production.

Purebred. A duck having parents of the same breed.

Ration. The feed consumed during the course of a day.

Roached back. A deformed, twisted, or humped back.

Scoop-bill. A bill with a top line that is severely concaved.

Secondaries. The flight feathers growing on the inner half of the wing.

Sex feathers. The curled feathers in the center of the tail when Mallard or Mallard-derived drakes are in nuptial plumage.

Sport. A colloquial term for mutation.

Standard-bred. Ducks that have been stringently selected over many generations according to the ideal that is set forth in the *Standard of Perfection*.

Standard of Perfection. A book containing pictures and descriptions of the physical characteristics desired in the perfect bird of each recognized breed and variety of poultry.

Starting ration. A high-protein feed used the first couple of weeks to get ducklings off to a good start.

Straight run. Young poultry that have not been sexed.

Strain. A group of animals within a breed that are more closely related than the general population of that breed.

Strain-cross. The mating together of males from one strain to females of a second strain.

Tertials. The band of large wing feathers growing next to the body.

Trio. A breeding unit consisting of a male and two females.

Type. The unique shape and size shared by animals of the same breed.

Variety. A subdivision of the breeds. In ducks, the varieties within a breed are identified by their plumage color or markings.

Waterfowl. Birds that naturally spend most of their lives on and near water. This term is often used in specific reference to ducks, geese, and swans.

Young drake or duck. A drake or duck less than one year old.

INDEX

Note: Page numbers in *italics* indicate illustrations; those in **boldface** indicate charts or tables.

Metric Conversion Chart

Unit	To convert to	Multiply it by
inches	centimeters	2.54
feet	meters	0.305
square feet	square meters	0.092
ounces	grams	28.35
pounds	kilograms	0.45
teaspoons	milliliters	4.93
tablespoons	milliliters	14.79
cups	milliliter	236.59
quarts	liters	0.95
gallons	liters	3.78
pounds	kilograms	0.45

Note: To convert Fahrenheit to Celsius, subtract 32 from the Fahrenheit number. Divide that number by 9. Multiply the answer by 5.

OTHER STOREY TITLES YOU WILL ENJOY

Basic Butchering of Livestock & Game by John J. Mettler, Jr., DVM. Provides clear, concise, step-by-step information for individuals who wish to slaughter their own meat. 208 pages. Paperback. ISBN 0-88266-391-7.

Building Small Barns, Sheds & Shelters by Monte Burch. Covers tools, materials, foundations, framing, sheathing, wiring, plumbing, and finish work for barns, woodsheds, garages, fencing, and animal housing. 248 pages. Paperback. ISBN 0-88266-245-7.

A Guide to Canning, Freezing, Curing and Smoking Meat, Fish and Game by Wilbur F. Eastman, Jr. Safe, step-by-step instructions for preparing and storing fresh meat, plus recipes and instructions for building smokehouses. 240 pages. Paperback. ISBN 1-58017-457-4.

Fences for Pasture & Garden by Gail Damerow. The complete guide to choosing, planning, and building today's best fences: wire, rail, electric, high-tension, temporary, woven, and snow. 160 pages. Paperback. ISBN 0-88266-753-X.

How to Build Small Barns & Outbuildings by Monte Burch. This book takes the mystery out of small-scale construction. Projects are offered with complete plans and instructions. 288 pages. Paperback. ISBN 0-88266-773-4.

Making Your Small Farm Profitable by Ron Macher. This practical, step-by-step guide to operating a small farm examines 20 alternative farming enterprises. Readers will learn how to target niche markets and sustain a farm's biological and economic health. 288 pages. Paperback. ISBN 1-58017-161-3.

Storey's Guide to Raising Poultry by Leonard S. Mercia. Provides current and up-to-date information on selecting birds for meat or egg production; chickens, turkeys, ducks, geese, and game birds; housing and equipment; brooding and rearing; home processing of eggs and poultry; and flock health. 352 pages. Paperback. ISBN 1-58017-263-6.

Wild Turkeys: Hunting and Watching by John J. Mettler, Jr., DVM. This book explains how to choose clothing, firearms, and equipment; turkey distribution, feeding patterns, and mating rituals; and how to dress a turkey carcass, mount a prize bird, and photograph turkeys in the wild. 176 pages. Paperback. ISBN 1-58017-069-2.

These and other books from Storey Publishing are available wherever quality books are sold or by calling 1-800-441-5700. Visit us at www.storey.com.